Springer Theses

Recognizing Outstanding Ph.D. Research

Aims and Scope

The series "Springer Theses" brings together a selection of the very best Ph.D. theses from around the world and across the physical sciences. Nominated and endorsed by two recognized specialists, each published volume has been selected for its scientific excellence and the high impact of its contents for the pertinent field of research. For greater accessibility to non-specialists, the published versions include an extended introduction, as well as a foreword by the student's supervisor explaining the special relevance of the work for the field. As a whole, the series will provide a valuable resource both for newcomers to the research fields described, and for other scientists seeking detailed background information on special questions. Finally, it provides an accredited documentation of the valuable contributions made by today's younger generation of scientists.

Theses are accepted into the series by invited nomination only and must fulfill all of the following criteria

- They must be written in good English.
- The topic should fall within the confines of Chemistry, Physics, Earth Sciences, Engineering and related interdisciplinary fields such as Materials, Nanoscience, Chemical Engineering, Complex Systems and Biophysics.
- The work reported in the thesis must represent a significant scientific advance.
- If the thesis includes previously published material, permission to reproduce this must be gained from the respective copyright holder.
- They must have been examined and passed during the 12 months prior to nomination.
- Each thesis should include a foreword by the supervisor outlining the significance of its content.
- The theses should have a clearly defined structure including an introduction accessible to scientists not expert in that particular field.

More information about this series at http://www.springer.com/series/8790

Roberto Navarro García

Predicting Flow-Induced Acoustics at Near-Stall Conditions in an Automotive Turbocharger Compressor

A Numerical Approach

Doctoral Thesis accepted by
Universitat Politècnica de València, Spain

 Springer

Author
Dr. Roberto Navarro García
CMT-Motores Térmicos
Universitat Politècnica de València
Valencia
Spain

Supervisor
Prof. José Galindo
CMT-Motores Térmicos
Universitat Politècnica de València
Valencia
Spain

ISSN 2190-5053   ISSN 2190-5061   (electronic)
Springer Theses
ISBN 978-3-319-72247-4   ISBN 978-3-319-72248-1   (eBook)
https://doi.org/10.1007/978-3-319-72248-1

Library of Congress Control Number: 2017961501

Printed on acid-free paper

This Springer imprint is published by Springer Nature
The registered company is Springer International Publishing AG
The registered company address is: Gewerbestrasse 11, 6330 Cham, Switzerland

Parts of this thesis have been published in the following journal articles:

Journal Papers

1. A. Broatch, J. Galindo, R. Navarro, and J. García-Tíscar. "Methodology for experimental validation of a CFD model for predicting noise generation in centrifugal compressors." *International Journal of Heat and Fluid Flow* 50 (2014), pp. 134–144. doi: 10.1016/j.ijheatfluidflow.2014.06.006
2. J. Galindo, A. Tiseira, R. Navarro, and M. López. "Influence of tip clearance on flow behavior and noise generation of centrifugal compressors in near-surge conditions." *International Journal of Heat and Fluid Flow* 52 (2015), pp. 129–139. doi: 10.1016/j.ijheatfluidflow.2014.12.004
3. A. Broatch, J. Galindo, R. Navarro, J. García-Tíscar, A. Daglish, and R. K. Sharma. "Simulations and measurements of automotive turbocharger compressor whoosh noise." *Engineering Applications of Computational Fluid Mechanics* 9(1) (2015). doi: 10.1080/19942060.2015.1004788
4. A. Broatch, J. Galindo, R. Navarro, and J. García-Tíscar. "Numerical and experimental analysis of automotive turbocharger compressor aeroacoustics at different operating conditions." *International Journal of Heat and Fluid Flow* (2016). doi: 10.1016/j.ijheatfluidflow.2016.04.003

Conference Papers

5. J. M. Desantes, A. Broatch, J. Galindo, and R. Navarro. "Aeroacustical and Flow Analysis in an Automotive Turbocharger Centrifugal Compressor by CFD Calculations and Measurements." In: 11th *European Turbomachinery Conference*. 2015

(...) it takes all the running you can do, to keep in the same place. If you want to get somewhere else, you must run at least twice as fast as that!.
The Red Queen, in "Carroll, Lewis. Through the Looking-Glass, and What Alice Found There *(1871)".*

Supervisor's Foreword

I am delighted to present to the reader of this Springer Theses Series the doctoral work carried out by Assistant Professor Roberto Navarro at CMT-Motores Térmicos Institute of the Universitat Politècnica de València in Spain. As supervisor of this Thesis, I have enjoyed during the four-year period the many qualities of Dr. Navarro, in particular, his determination for perfection and his attention to the details. It is not certainly the first time when the supervisor learns from his trainee.

The Thesis primary objective was the understanding of aeroacoustical effects that lead to the noisy behavior of turbocharger centrifugal compressors when operating near their surge limit. The context in which this objective became important was the design strategy chosen by carmakers to improve performance and emissions of automotive engines, both diesel and gasoline fueled. The downsizing, together with downspeeding, allows the automotive power plant to operate at the best efficiency and lowest emissions conditions. However, to achieve this, the centrifugal compressor requirements are augmented in terms of boosting pressure and flow range. With downsized engines, the turbocharger compressor is forced to operate at high compression ratio and low mass flow where the so-called whoosh noise becomes noticeable.

The reader will find in this volume a methodology mainly based on 3D CFD simulations supported with experiments of an automotive compressor in an anechoic chamber. The strategy for the simulations validation relies on the comparison of measured and calculated spectra of flow properties, which required the development of a novel signal processing methodology. Another interesting part of the Thesis was devoted to account for the importance of tip clearance and shaft motion on the whoosh generation. Finally, the analysis of the unsteady calculations showed an insight on the flow structures that can be on the root cause of whoosh.

The model and analysis strategy developed in this book have been the basis for subsequent research on the impact of intake geometry on overall compressor performance, generated noise, and surge margin. Besides, water condensation at the compressor inlet due the usage of low-pressure exhaust gas recirculation (EGR) has been investigated in combination with the model presented in these pages. As a matter of fact, a doctoral Thesis covering the aforementioned topics is currently being supervised by Roberto Navarro.

Valencia, Spain   Prof. José Galindo
October 2017

Acknowledgements

First of all, I would like to express my sincere gratitude to Prof. Payri and Prof. Desantes on behalf of CMT-Motores Térmicos and Universitat Politècnica de València for allowing me to become part of this research group and for providing me with the material and human resources required for the accomplishment of this doctoral thesis.

I owe my most special thanks to Prof. Galindo for his guidance and support. I have enjoyed the many interesting discussions held, ranging from turbomachinery flow to PowerPoint-related psychology! Many thanks to my officemates, co-workers, and colleagues Pablo Fajardo and Daniel Tarí, for their help in countless matters. Thanks to Prof. Broatch, Jorge García, and Bernardo Planells for the obtention and help in the discussion of experimental acoustic measurements. Also thanks to Andrés Tiseira and Miguel Andrés López for providing tip clearance measurements.

I would also like to thank Sergio Hoyas, Antonio Gil, and Pau Raga for their help regarding CFD meshing, postprocessing, and HPC resources management. Also thanks to UPV and Paco Rosich for use and support of Rigel cluster. I would like to give sincere thanks to the administrative staff of CMT-Motores Térmicos for helping me with all the paperwork. I would like to acknowledge also the rest of CMT colleagues for creating this nice work ambience and increasing my knowledge at several fields. To cite a few of them: Vicente Dolz, Pedro Martí, Marcos Carreres, etc.

I would like to thank Håkan Nilsson and the rest of the staff at the Division of Fluid Dynamics for being such good hosts during my doctoral stay at Chalmers University of Technology. I am deeply grateful to Leontina Di Cecco and the rest of Springer personnel for the support to publish this book. It is a great honor to contribute to the Springer Theses collection.

Last but not least, I want to thank my family and friends for their endless love and support despite I am always working around the clock. I really appreciate it. And thank you, Raquel, for making every day worth living.

Valencia, Spain
October 2017

Contents

Symbols

Latin Characters

a	Speed of sound ($m \cdot s^{-1}$)
c_p	Specific heat capacity at constant pressure ($J \cdot kg^{-1} \cdot K^{-1}$)
D	Duct diameter (m)
f	Frequency (Hz)
f_N	Nyquist frequency (Hz)
$\widetilde{f_d}$	IDDES blending parameter (–)
H	Helicity ($m \cdot s^{-2}$)
I	Sound intensity ($W \cdot s^{-2}$)
I	Turbulence intensity (%)
L	Sensor separation (m)
$\dot{m}$	Mass flow rate ($kg \cdot s^{-1}$)
m, s, p	Meridional, spanwise, or pitchwise coordinate, respectively (%)
M	Mach number (–)
N	Compressor rotational speed (rpm)
p	Pressure (Pa)
$\vec{r}$	Position vector (m)
r	Radial coordinate (m)
t	Time (s)
t	Parametric angle (rad)
T	Temperature (K)
u	Axial velocity ($m \cdot s^{-1}$)
u^+	Dimensionless velocity (–)
u_τ	Friction velocity ($m \cdot s^{-1}$)
$\vec{v}$	Velocity ($m \cdot s^{-1}$)
$\dot{W}$	Compressor absorbed power ($kg \cdot m^2 \cdot s^{-3}$)
W_u	Compressor specific work ($m^2 \cdot s^{-2}$)

y^+	Dimensionless wall distance (–)
z	Axial coordinate (m)

Greek Characters

α	Velocity angle (°)
η	Efficiency (%)
ε	Relative difference (%)
γ	Ratio of specific heats (–)
ν	Kinematic viscosity ($\mathrm{m^2 \cdot s^{-1}}$)
ϕ	Generic variable
ϕ	DDES correction factor (–)
$\Pi_{t,t}$	Total-to-total pressure ratio (–)
ρ	Density ($\mathrm{kg \cdot m^{-3}}$)
τ	Compressor torque ($\mathrm{kg \cdot m^2 \cdot s^{-2}}$)
θ	Angular coordinate (rad)
$\vec{\omega}$	Rotational speed ($\mathrm{rad \cdot s^{-1}}$)

Sub- and Superscripts

$*$	Corrected variable
0	Stagnation variable
$\bar{}$	mean variable
$1, 2, 3$	1st, 2nd, or 3rd probe in the array, respectively, numbered in downstream order
$1/3, 2/3, 3/3$	Related to one-third, two-thirds, or three-thirds of baseline clearance, respectively
a	Related to asymmetric mode
$back$	Backward traveling wave
$base$	Related to baseline case
c	Cutoff (frequency)
c	Related to an ellipse center
CFD	Related to simulation
exp	Related to experimental measurement
in	Inlet duct
$forw$	Forward traveling wave
i	Intrinsic (of the array)
out	Outlet duct
r	Relative
r	Related to radial mode
r, θ, z	(Vector) cylindrical components
rad	Related to radial direction
s	Isentropic

s	Related to a certain span
ref	Reference value
$tang$	Related to tangential direction
W, A, S	Wall, axis, or cross-sectional monitor, respectively
x, y, z	(vector) Cartesian components

Acronyms

0D	Zero dimensional
1D	One dimensional
3D	Three dimensional
BC	Boundary condition
BEP	Best efficiency point
BL	Boundary layer
BPF	Blade-passing frequency
CAA	Computational aeroacoustics
CAD	Computer-aided design
CBV	Compressor bypass valve
CFD	Computational fluid dynamics
CFL	Courant–Friedrich–Levy (number)
DDES	Delayed DES
DES	Detached eddy simulation
DMD	Dynamic mode decomposition
DNS	Direct numerical simulation
FEM	Finite element method
IDDES	Improved delayed DES
LCMV	Linearly constrained minimum variance
LES	Large eddy simulation
LIC	Line integral convolution
LLM	Log-layer mismatch
MoC	Method of Characteristics
MRF	Multiple reference frame
NRBC	Non-reflecting BC
NVH	Noise, vibration, and harshness
PIV	Particle image velocimetry
POD	Proper orthogonal decomposition
PS	(blade) Pressure side
PSD	Power spectral density
RANS	Reynolds-averaged Navier–Stokes
RBM	Rigid body motion
rpm	Revolutions per minute
SGS	Sub-grid scale
SIL	Sound intensity level

SPL	Sound pressure level
SS	(blade) Suction side
TCN	Tip clearance noise
TCR	Tip clearance ratio
URANS	Unsteady RANS
WMLES	Wall-modeled LES

Chapter 1
Introduction

1.1 Motivation

The use of turbocharging is almost as old as the internal combustion engine itself. Büchi introduced in the first decade of the twentieth century this technology, increasing the rated power of engines through the density rise obtained by a turbocompressor. Turbochargers were first and progressively introduced in aeronautical reciprocating engines, which suffered from a decline in power due to the decrease of the air density when an aircraft would fly at high altitude. Turbocharging was a way to overcome this limitation. Turbocharging was then adopted by large displacement Diesel engines as those used in marine propulsion, locomotives or stationary powerplants. The pairing of Diesel engines and turbocharging technology has been a successful story since then, given that the increase of intake pressure and temperature that the supercharger provides improves the conditions for the development of Diesel combustion. This lead to an increase, not only in power, but also in thermal efficiency.

The introduction of turbocharging in automotive engines was quite more recent. In the eighties, turbocharged petrol engines were used successfully in Formula 1. The increase in power was so impressive that they were finally banned. The use in passenger cars comes from the same period. The objective was again to improve the rated power and to reduce fuel consumption. However, these first automotive turbocharged engines suffered from the so-called turbocharger lag, a delay in the throttle response due to mechanical and thermal inertia of a too-large turbocharger. It was in the nineties when the turbocharger became essential in automotive Diesel engines. The introduction of more and more stringent pollutant emissions regulations made necessary a better control of fuel and air flows. The first was accomplished with new electronically controlled injections systems. The control of air was achieved by waste-gated and variable geometry turbochargers. The latter, in turn, solved the problem of turbocharger lag. It is worth to mention here that the main method to reduce nitrogen oxide emissions was the Exhaust Gas Recirculation. The only way to introduce exhaust gas into the cylinders without penalizing the oxygen content is to boost the fresh charge at a greater pressure.

© Springer International Publishing AG 2018
R. Navarro García, *Predicting Flow-Induced Acoustics at Near-Stall Conditions in an Automotive Turbocharger Compressor*, Springer Theses, https://doi.org/10.1007/978-3-319-72248-1_1

Today, turbocharging does not mean power, but lower emissions and fuel consumption [1]. Actually, the increase in power density yielded by the turbocharger is used to decrease the cylinder displacement and speed so that the engine works in a different operating point for the same driving conditions, with higher efficiency and lower emissions. This is known as downsizing (decrease in displacement) and downspeeding (decrease in speed) techniques, which have been the base of the improvement of automotive Diesel engines in the last decade, together with other techniques such as fuel injection and aftertreatment [2]. This downsizing-downspeeding technique is being implemented in the last years in petrol engines as well [3, 4].

However, turbocharging downsized and downspeeded engines arises new challenges for the turbomachinery. In the turbine side, an increase of gas temperature has been allowed by the use of new materials. Flow pulsation amplitude at the turbine inlet has increased too, especially when the downsizing is carried out reducing the cylinders count. But probably the biggest challenges are in the compressor side. Downsizing and downspeeding demand from the compressor high compression ratio for the lowest possible flow and at the same time enough flow capacity for the rated power condition. It is known that after the introduction of Euro V norm, several car makers suffered from compressor surge. Therefore, the most relevant development today in turbocharger compressors is the increase of available flow range from surge to choke while maintaining the efficiency. Compressor designs have evolved to a reduced trim in order to increase the flow coefficient. Also, impellers with backswept blades are used to improve flow stability.

An important issue related to the use of turbomachinery in internal combustion engines is its contribution to noise emissions. Braun et al. [5] performed a literature review of noise source characteristics in the ISO 362 vehicle pass-by noise test. They found that noise source ranking varies depending on the author in question, but intake and exhaust systems are regarded as one of the main noise sources. Moreover, Stoffels and Schroeer [6] showed that radiated noise in low speed range (below 2800 rpm) of a downsized turbocharged gasoline powertrain is higher than that of an engine with the same power but larger displacement.

Up until now, turbocharger noise has been somehow hidden by the engine noise, but the latter has been significantly decreased in the last years. For instance, great progress has been achieved in reducing the combustion noise through the shaping of injection rate with electronically controlled injection systems [7]. The turbocharger is indeed responsible for the attenuation of intake and exhaust noise through the reduction of pressure pulsations or the increase of boost pressure [8]. This has been recently discussed with the re-introduction of turbocharger technology in Formula 1 competition, since the acoustic signature of the racing cars has been modified. Blade passing frequency (BPF) tonal noise and other high-frequency content have been known by researchers and addressed with the aid of silencers [9, 10]. However, strong downsizing of engines and increase of low speed torque seem to have raised an issue with turbocharger airborne noise [11], since compressor operating points are shifted towards surge region [12].

In particular, a broadband noise is detected during some engine conditions (full load, tip-in and tip-out maneuvers). Researchers refer to this noise with terms such as *hiss* [13] or *whoosh* [11, 12]. Whoosh noise is commonly described as a broadband noise either in the 1–3 kHz band [11, 14] or in frequencies ranging from 4 to 12 kHz [12, 15]. This noise is more audible when the compressor is working in near-surge conditions [11].

Gaudé et al. [13] analyzed different turbocharger related noises and their sources. They identified tonal noises at sub-synchronous speed (oil whirl), synchronous speed and harmonics (due to rotating unbalance or lack of symmetry of turbocharger rotating assembly) and super-synchronous speed (turbine and compressor BPF). Compressor hiss noise was regarded as the only broadband noise mechanism. For a thorough and detailed description of the different noise sources in automotive turbochargers, please refer to the work of Nguyen-Schäfer [16].

According to the literature review, airborne turbocharger noise is a topic of major concern for automotive manufacturers. However, there is no agreement on which are the compressor working conditions with highest noise emissions or which is the acoustic signature of these points, particularly when whoosh noise phenomenon is involved.

1.2 Background

To the respondant's knowledge, the most extensive experimental collection of papers regarding centrifugal compressor aeroacoustics is the one produced by researchers of Pennsylvania State University. The papers of Mongeau et al. [17], Choi [18], Mongeau et al. [19] and Choi et al. [20] are an excerpt of the aforementioned collection. They conducted experimental studies on a centrifugal pump with no diffuser, so that the turbomachine discharged to the atmosphere in an anechoich chamber. However, the differences in impeller geometry and speed between the studied centrifugal pump (upper limit of 3600 rpm) and modern turbocharger compressors (which can spin well above 200 krpm), makes it difficult to take advantage of the knowledge provided by these works. For instance, their frequency range of interest was established between 100 and 800 Hz [18], not even covering the rotation order related to the compressor speed studied in this book (160 krpm).

Raitor and Neise [21] conducted an experimental study to describe the sound generation mechanisms of centrifugal compressors. At compressor design speed, with supersonic flow conditions, the main features are blade tone noise and buzz-saw noise. For low compressor speeds, corresponding to subsonic flow conditions, radial compressor noise is dominated by tip clearance noise (TCN). TCN is a narrow-band noise observed at frequencies about half the BPF, which increase with speed. For their operating conditions, TCN ranged from 3 to 4 kHz. Raitor and Neise concluded that TCN is due to the secondary flow through the gap between rotor blade tips and the casing wall. They did not include any comment on whoosh noise in their paper, so it is not clear beforehand whether whoosh noise is related to TCN.

The centrifugal compressor studied by Raitor and Neise [21] is supposed to belong to a turbocharger, but there still exists a difference in speed with small turbocharger compressors (maximum compressor speed investigated by Raitor and Neise was 50 krpm). They studied mainly supersynchronous phenomena, such as blade passing tone and TCN. In this work, though, whoosh noise is the major concern, which is regarded as a subsynchronous phenomenon by several researchers [11, 14]. The main difference between the compressor studied by Raitor and Neise and passenger car turbocharger compressors such as the one studied in this work is that, for the latter, human peak hearing sensitivity lays close to compressor rotation order.

The aforementioned works by Evans and Ward [11], Teng and Homco [12] or Sevginer et al. [22] dealt with small turbocharger compressor NVH issues by measuring noise radiation. However, their approach corresponded to research performed in automotive brand manufacturers, i.e., they encountered a problem (whoosh noise) at certain operating conditions and they focused on solving the issue. A common way to reduce noise radiation was the use of resonators in compressor outlet hose [11, 12, 22]. Nevertheless, if flow phenomena leading to whoosh noise were known, quieter centrifugal compressors could be designed. For instance, some of the presented works [11–13, 22] postulate turbulence generation as the source of whoosh noise, but no evidence supporting this assertion is provided. Numerical simulations could be used as a tool to acquire knowledge about whoosh noise phenomenon.

Mendonça et al. [23] analyzed flow-induced aeroacoustics of an automotive radial compressor using Detached Eddy Simulations (DES). A narrow band noise at a frequency about 70% of rotational speed was detected, corresponding to 2.5 kHz for the investigated compressor speed. Rotating stall was found to be the source of the narrow band noise. Mendonça et al. did not include experimental measurements in their work. Therefore, it can not be confirmed whether the narrow band attributed to rotating stall corresponds to whoosh noise.

Karim et al. [24] and Lee et al. [25] combined numerical simulations and experiments to study compressor noise radiation. However, Karim et al. only used measurements to confirm the reduction of overall radiated whoosh noise level when increasing the length of the leading edge step. Lee et al. compared measured and numerical prediction of compressor noise spectrum, showing a lack of agreement. Probably, the calculation of only one revolution data for acoustic analogy led to poor spectrum estimation.

It is obvious that the state of the art conditions the scope of any research, but previous in-house know-how constrains it even more. At CMT-Motores Térmicos, CFD research on radial turbomachinery started less than a decade ago.

Margot et al. [26] analyzed the flow in a centrifugal compressor near the surge line using CFD. In their transient calculations, unstable behavior appeared when approaching the surge line. In these conditions, there are zones located upstream of the inducer with low pressure levels, that lead to reverse flow in the areas of the periphery of the impeller placed immediately downstream of these regions.

Lang [27] performed unsteady CFD calculations of a centrifugal compressor from a stable operating point until deep surge. A 0D plenum volume model was attached to the outlet boundary condition by means of a User Defined Function in FLUENT. The model simulates the outlet ducts and throttle valve existing in the experimental facility. The throttle valve area was decreased in the model to drive the compressor to surge. Surge cycle was studied and validated against experimental results and 1D computations. Flow was analyzed in different compressor conditions (stable, near surge, maximum negative mass flow, etc.). A parametric study concerning the influence of inlet geometries (straight duct, elbow and plenum) into surge margin was also performed. An experimental campaign was undertaken to obtain the surge line of the different configurations, while 3D numerical computations provided some insight into the flow behavior.

Fajardo [28] conducted a numerical study on a variable nozzle turbine. A comparison between turbine global variables with different setup parameters (such as mesh motion strategy or solver type) and experimental measurements was carried out. A local mesh independence study was undertaken, obtaining an almost optimal grid density [29]. Finally, the turbine was simulated under engine-like pulsating conditions. Some hysteretic effects were noticed such as detachment and reattachment of the flow in the nozzle vanes. Quasi-steady assumption was thus violated and a simplified model consisting of several lumped element models corresponding to the different turbine regions (volute, nozzle and rotor) was developed [30].

Besides, the previous know-how of pressure decomposition techniques has been a determining factor for the development of this work. There is a vast bibliography of in-house papers and theses dealing with this topic, but it is worth highlighting the papers that have used satisfactorily pressure wave decomposition to analyze gas dynamics in intake and exhaust systems of turbocharged ICEs by means of 1D codes (see, for instance, [31–34]). Fajardo [28] developed a 1D–3D coupling through the use of Riemann decomposition as well (see [35–37]).

To conclude this section, it seems appropriate to quote the sentence stated by Cumpsty in 1977 [38]: "(…) the literature seems to indicate that there has been no research on high speed radial compressor noise". 40 years later, the knowledge of high speed centrifugal compressor aeroacoustics is much greater, but it feels that there is plenty of room for improvement.

1.3 Main Objectives

The aim of this book is to contribute to the understanding of turbocharger compressor flow-induced acoustics. This can be expressed in the form of specific individual objectives:

- Understand which is the flow phenomena leading to whoosh noise.
- Analyze other significant compressor aeroacoustic features that are present in human hearing range (20–20,000 Hz).
- Determine the evolution of compressor acoustic signature with operating conditions.
- Analyze the change of compressor flow behavior as the working point is shifted towards surge.

1.4 Approaches to Compute Flow-Induced Noise

As stated in Sect. 1.2, numerical simulations of centrifugal compressor flow would be helpful to gain insight of the whoosh noise phenomenon, which is the main objective of this work. There are several approaches to perform Computational Aeroacoustics (CAA) and, according to Nogueira et al. [39], "(…) there is still no consensus about the aeroacoustic approach to adopt, and actually, it depends on the application". In this section, a brief description of three of the most popular methods to compute aerodynamically generated noise is provided. The Direct Method, Integral Method Based on Acoustic Analogy and Broadband Noise Source Models will be covered.

1.4.1 Direct Method

With this approach, both generation and propagation of sound waves are directly computed by solving the unsteady compressible Navier-Stokes turbulent equations. The direct method is computationally expensive because it also requires high-order discretization schemes in time and space and very fine computational meshes all the way to the acoustic probes. Accurate resolution of the propagation of acoustic waves imposes time step restrictions such that the Courant number is in the range of 1–2 at most. The computational cost becomes prohibitive when sound is to be predicted in the far field, e.g., hundreds of chord-lengths in the case of an airfoil. The direct method is only feasible when receivers are in the near field.

For instance, Mendonça et al. [23] used the direct method to compute the noise of a turbocharger compressor inside inlet and outlet pipes.

1.4.2 Integral Method Based on Acoustic Analogy

For predictions of mid- to far-field noise, the methods based on Lighthill's acoustic analogy offer viable alternatives to the direct method. The acoustic analogy essentially decouples the propagation of sound from its generation. The near-field flow is obtained from appropriate governing equations. It is important to highlight that RANS formulation does not provide any information on turbulent flow structures and spectral distribution, which might be of importance to predict flow-induced noise. Therefore, Large Eddy Simulations (LES) or DES should be used when the broadband contribution is significant.

The acoustic analogy predicts sound generated by equivalent acoustic sources obtained from surface integrals and volume integrals. Therefore, time-accurate solutions of the flow-field variables, such as pressure, velocity components, and density on source (emission) surfaces, are required to evaluate the integrals. Using this technique, time histories of acoustic signals at prescribed receiver locations can be computed. Both broadband and tonal noise can be predicted depending on the nature of the flow (noise source) being considered, the turbulence model employed, and the time scale of the flow resolved in the flow calculation.

The acoustic analogy is applicable only to predicting the propagation of sound toward free space, but not for internal flows, since the vibrating ducts are the "sources" regarding the transmission to the outside. Prediction of noise using this approach for a confined, internal flow applications will be of limited usefulness because the flow tends to be more strongly coupled with the acoustics.

1.4.3 Broadband Noise Source Models

In many practical applications involving turbulent flows, noise does not have any distinct tones, and the sound energy is continuously distributed over a broad range of frequencies. This is not the case of turbomachinery, in which there are frequency peaks related to blade passing. However, broadband models are still useful to detect noise sources. Statistical turbulence quantities readily computable from RANS equations can be utilized, in conjunction with semi-empirical correlations and Lighthill's acoustic analogy, to shed some light on the source of broadband noise. The source models can be employed to extract useful diagnostics on the noise source to determine which portion of the flow is primarily responsible for the noise generation. However, these source models do not predict the sound at receivers. The broadband noise source models do not require transient solutions to any governing fluid dynamics equations, thus being the least expensive approach to computing aerodynamically generated noise.

Li et al. [40] employed a broadband noise source model to evaluate compressor noise when different types of inlet geometries and impeller blades are used.

1.4.4 Selected Approach

The objectives stated in Sect. 1.3 are pursued in this book by performing 3-dimensional CFD simulations of a complete model of a turbocharger compressor, including part of the inlet and outlet pipes so that the direct method described in Sect. 1.4.1 can be applied. These simulations are carried out with the commercial code StarCCM+ of CD-Adapco. This code is based on a finite-volume approach, with 2nd order discretizations in time and space. During the framework of this work, StarCCM+ versions from 7.02.011 [41] to 9.02.005 [42] have been used, all of them with double precision to avoid issues with propagation of acoustic waves [43].

1.5 Methodological Objectives

When CFD is defined as the tool to achieve the goals defined in Sect. 1.3, additional issues arise.

As stated in Sect. 1.2, turbocharger compressor CAA is a recent topic which has been covered by not many researchers. Therefore the influence of the numerical setup on compressor noise prediction should be investigated in this work.

Moreover, some works [21, 23, 44] consider tip leakage flow as a key point in the stability and noise production of centrifugal compressors. With the current computational capabilities, tip clearance can be modeled, but tip clearance profile relies on the CAD files that would not reflect tip clearance reduction due to the actual compressor working conditions. It is thus required to analyze the sensitivity of compressor noise to tip clearance.

In any case, experimental measurements are required to assess the ability of these simulations to capture compressor noise generation. However, the computational cost of modeling the whole turbocharger test rig is increased because of the existence of long ducts [37]. It is a common approach to set virtual pipes' length so as to allow the development of the flow from boundary conditions instead of using the actual duct length. The methodology to compare numerical and experimental spectra that will be not obtained at the same location is thus a key point to obtain a meaningful comparison.

The first issue of this work arises at this early stage. No conclusions about best numerical configuration can be obtained until a suitable methodology to compare experimental measurements and numerical predictions is developed. However, a setup is required to obtain such CFD results that allows one to define the proper method to perform the validation of the model. This feedback loop is broken by setting a baseline configuration, that may not be optimal, to be able to develop the validation methodology. Once a suitable means of comparison is defined, the baseline setup will be analyzed. In this book, the numerical configuration of Mendonça et al. [23] will be used as a reference to define the baseline model.

To sum up, the present work should accomplish the following methodological objectives:

• Develop a method to compare numerical and experimental spectra, which are obtained at different locations.
• Analyze the influence of tip clearance on compressor noise generation.
• Investigate the sensitivity of compressor noise to main setup parameters.
• Find or develop adequate postprocessing tools so as to analyze radial turbomachinery flow and aeroacoustics.

1.6 Book Outline

The remainder of this work is organized in the following manner.

Chapter 2 presents the experimental measurements that will be used as a means of validation for CFD predictions. The numerical model, which at this stage is strongly influenced by the work of Mendonça et al. [23], will be described. The methodology to compare CFD and experimental spectra will be developed.

Chapter 3 covers the estimation of tip clearance reduction during actual compressor operation. Using these values, compressor noise sensitivity to tip clearance will be assessed, followed by an investigation of flow in the vicinity of tip clearance.

Chapter 4 is devoted to the analysis of the influence of most important set-up parameters in both compressor global variables and generation of noise. Wheel rotation strategies and turbulence models will be discussed. Then, sensitivity analysis to grid density, solver type and time-step size will be performed so as to judge the baseline configuration inspired by the work of Mendonça et al. [23].

Chapter 5 deals with the description of the main features of the time-averaged flow. Three working points at same compressor speed will be studied, ranging from best efficiency point to near surge conditions.

Chapter 6 provides a description of the changes of acoustic signature as working point is moved towards surge. The aeroacoustics of the three working points will be analyzed in terms of temporal evolution of the flow field so as to justify the features observed in the pressure spectra, with particular interest on whoosh noise phenomenon.

Finally, Chap. 7 incorporates a summary of the findings and contributions of this work. Based on these findings and the issues encountered, recommendations for future work are also included.

Besides, Appendix A explains different postprocessing tools developed to analyze radial turbomachinery flow and aeroacoustics.

References

1. W. Knecht, Diesel engine development in view of reduced emission standards. Energy **33**(2), 264–271 (2008). https://doi.org/10.1016/j.energy.2007.10.003
2. T.V. Johnson, Review of diesel emissions and control. Int. J. Engine Res. **10**(5), 275–285 (2009). https://doi.org/10.1243/14680874JER04009
3. T. Lake, J. Stokes, R. Murphy, R. Osborne, A. Schamel, *Turbocharging Concepts for Downsized DI Gasoline Engines*. SAE Technical Paper (2004). https://doi.org/10.4271/2004-01-0036
4. W. Bandel, G.K. Fraidl, P.E. Kapus, H. Sikinger, C. Cowland, *The Turbocharged GDI Engine: Boosted Synergies for High Fuel Economy Plus Ultra-low Emission*. SAE Technical Paper (2006). https://doi.org/10.4271/2006-01-1266
5. M. Braun, S. Walsh, J. Horner, R. Chuter, Noise source characteristics in the ISO 362 vehicle pass-by noise test: literature review. Appl. Acoust. **74**(11), 1241–1265 (2013). https://doi.org/10.1016/j.apacoust.2013.04.005. (ISSN:0003-682X)
6. H. Stoffels, M. Schroeer, *NVH Aspects of a Downsized Turbocharged Gasoline Powertrain with Direct Injection*. SAE Technical Paper 2003-01-1664 (2003). https://doi.org/10.4271/2003-01-1664
7. M. Badami, F. Mallamo, F. Millo, E. Rossi, *Influence of Multiple Injection Strategies on Emissions, Combustion Noise and BSFC of a DI Common Rail Diesel Engine*. SAE Technical Paper (2002). https://doi.org/10.4271/2002-01-0503
8. A.J. Torregrosa, A. Broatch, X. Margot, J. Gómez-Soriano, Towards a predictive CFD approach for assessing noise in diesel compression ignition engines, in *The Ninth International Conference on Modeling and Diagnostics for Advanced Engine Systems (COMODIA 2017)* (2017)

9. A.J. Torregrosa, A. Broatch, V. Raimbault, J. Migaud, Compact high-pressure intake silencer with multilayer porous material. SAE Int. J. Passeng. Cars-Mech. Syst. **9**(3), 1078–1085 (2016). https://doi.org/10.4271/2016-01-1819

10. L. Chen, C. Yipeng, Z. Wenping, Z. Xiaochen, S. Hua, An intake silencer for the control of marine diesel turbocharger compressor noise, in *24th International Congress of Sound and Vibration* (London, 2017)

11. D. Evans, A. Ward, *Minimizing Turbocharger Whoosh Noise for Diesel Powertrains*. SAE Technical Paper 2005-01-2485 (2005). https://doi.org/10.4271/2005-01-2485

12. C. Teng, S. Homco, Investigation of compressor whoosh noise in automotive turbochargers. SAE Int. J. Passeng. Cars-Mech. Syst. **2**(1), 1345–1351 (2009). https://doi.org/10.4271/2009-01-2053

13. G. Gaudé, T. Leèfvre, R. Tanna, K. Jin, T.J.B. McKitterick, S. Armenio, Experimental and computational challenges in the quantification of turbocharger vibro-acoustic sources, in *Proceedings of the 37th International Congress and Exposition on Noise Control Engineering*, vol. 2008, No. 3. (Institute of Noise Control Engineering, 2008) pp. 5598–5611. ISBN:978-1-60560-989-8

14. E.P. Trochon, *A New Type of Silencers for Turbocharger Noise Control*. SAE Technical Paper **110**(6), 1587–1592 (2001). https://doi.org/10.4271/2001-01-1436

15. N. Figurella, R. Dehner, A. Selamet, K. Tallio, K. Miazgowicz, R. Wade, Noise at the mid to high flow range of a turbocharger compressor, in *41st International Congress and Exposition on Noise Control Engineering*, vol. 2012, No. 3 (Institute of Noise Control Engineering, 2012) pp. 786–797. ISBN:978-1-62748-560-9, http://www.ingentaconnect.com/content/ince/incecp/2012/00002012/00000003/art00015

16. H. Nguyen-Schäfer, *Aero and Vibroacoustics of Automotive Turbochargers* (Springer Science & Business Media, Berlin, 2013). https://doi.org/10.1007/978-3-642-35070-2

17. L. Mongeau, D. Thompson, D. McLaughlin, Sound generation by rotating stall in centrifugal turbomachines. J. Sound Vib. **163**(1), 1–30 (1993). https://doi.org/10.1006/jsvi.1993.1145

18. J.-S. Choi, Aerodynamic noise generation in centrifugal turbomachinery. KSME J. **8**(2), 161–174 (1994). https://doi.org/10.1007/BF02953265

19. L. Mongeau, D. Thompson, D. Mclaughlin, A method for characterizing aerodynamic sound sources in turbomachines. J. Sound Vib. **181**(3), 369–389 (1995). https://doi.org/10.1006/jsvi.1995.0146

20. J.-S. Choi, D.K. McLaughlin, D.E. Thompson, Experiments on the unsteady flow field and noise generation in a centrifugal pump impeller. J. Sound Vib. **263**(3), 493–514 (2003). https://doi.org/10.1016/S0022-460X(02)01061-1

21. T. Raitor, W. Neise, Sound generation in centrifugal compressors. J. Sound Vib. **314**, 738–756 (2008). https://doi.org/10.1016/j.jsv.2008.01.034. (ISSN: 0022-460X)

22. C. Sevginer, M. Arslan, N. Sonmez, S. Yilmaz, Investigation of turbocharger related whoosh and air blow noise in a diesel powertrain, in *Proceedings of the 36th International Congress and Exposition on Noise Control Engineering*, pp. 476–485. ISBN:978-1-60560-385-8 (2007)

23. F. Mendonça, O. Baris, G. Capon, Simulation of radial compressor aeroacoustics using CFD, in *Proceedings of ASME Turbo Expo 2012*. GT2012-70028. ASME. 2012, pp. 1823–1832. https://doi.org/10.1115/GT2012-70028

24. A. Karim, K. Miazgowicz, B. Lizotte, A. Zouani, *Computational Aero-Acoustics Simulation of Compressor Whoosh Noise in Automotive Turbochargers*. SAE Technical Paper 2013-01-1880 (2013). https://doi.org/10.4271/2013-01-1880

25. Y. Lee, D. Lee, Y. So, D. Chung, *Control of Airflow Noise from Diesel Engine Turbocharger*. SAE Technical Paper 2011-01-0933 (2011). https://doi.org/10.4271/2011-01-0933

26. X. Margot, A. Gil, A. Tiseira, R. Lang, *Combination of CFD and Experimental Techniques to Investigate the Flow in Centrifugal Compressors Near the Surge Line*. SAE Technical Paper 2008-01-0300 (2008), p. 7. https://doi.org/10.4271/2008-01-0300

27. R. Lang, *Contribución a la Mejora del Margen de Bombeo en Compresores Centrífugos de Sobrealimentación*. Ph.D. thesis. Universitat Politècnica de València, 2011, http://hdl.handle.net/10251/12331

28. P. Fajardo, *Methodology for the Numerical Characterization of a Radial Turbine Under Steady and Pulsating Flow*. Ph.D. thesis. Universitat Politècnica de València, 2012, http://hdl.handle.net/10251/16878
29. J. Galindo, S. Hoyas, P. Fajardo, R. Navarro, Set-up analysis and optimization of CFD simulations for radial turbines. Eng. Appl. Comput. Fluid Mech. **7**(4), 441–460 (2013). https://doi.org/10.1080/19942060.2013.11015484
30. J. Galindo, P. Fajardo, R. Navarro, L.M. Garíca-Cuevas, Characterization of a radial turbocharger turbine in pulsating flow by means of CFD and its application to engine modeling. Appl. Energy **103**, 116–127 (2013). https://doi.org/10.1016/j.apenergy.2012.09.013
31. A.J. Torregrosa, J.R. Serrano, J. Dopazo, S. Soltani, *Experiments on Wave Transmission and Reflection by Turbochargers in Engine Operating Conditions*. SAE Technical Paper 2006-01-0022 (2006). https://doi.org/10.4271/2006-01-0022
32. J.R. Serrano, F.J. Arnau, V. Dolz, A. Tiseira, C. Cervelló, A model of turbocharger radial turbines appropriate to be used in zeroand one-dimensional gas dynamics codes for internal combustion engines modelling. Energy Convers. Manage. **49**(12), 3729–3745 (2008). https://doi.org/10.1016/j.enconman.2008.06.031
33. A.J. Torregrosa, J. Galindo, J.R. Serrano, A. Tiseira, A procedure for the unsteady characterization of turbochargers in reciprocating internal combustion engines, in *Fluid Machinery and Fluid Mechanics* (Springer, Berlin, 2009), pp. 72–79
34. J.R. Serrano, F.J. Arnau, R. Novella, M.Á. Reyes-Belmonte, *A Procedure to Achieve 1D Predictive Modeling of Turbochargers Under Hot and Pulsating Flow Conditions at the Turbine Inlet*. SAE Technical Paper 2014-01-1080 (2014), 13pp. https://doi.org/10.4271/2014-01-1080
35. J. Galindo, A. Tiseira, P. Fajardo, R. Navarro, Coupling methodology of 1D finite difference and 3D finite volume CFD codes based on the method of characteristics. Math. Comput. Model. **54**(7–8), 1738 (2011). https://doi.org/10.1016/j.mcm.2010.11.078. (Mathematical Models of Addictive Behaviour, Medicine & Engineering, pp. 1738–1746. ISSN:0895-7177)
36. A.J. Torregrosa, P. Fajardo, A. Gil, R. Navarro, Development of a non-reflecting boundary condition for application in 3D computational fluid dynamic codes. Eng. Appl. Comput. Fluid Mech. **6**(3), 447–460 (2012). https://doi.org/10.1080/19942060.2012.11015434
37. J. Galindo, A. Tiseira, P. Fajardo, R. Navarro, Analysis of the influence of different real flow effects on computational fluid dynamics boundary conditions based on the method of characteristics. Math. Comput. Model. **57**(7–8), 1957 (2013). https://doi.org/10.1016/j.mcm.2012.01.016. (Public Key Services and Infrastructures EUROPKI-2010-Mathematical Modelling in Engineering & Human Behaviour 2011, pp. 1957–1964. ISSN:0895-7177)
38. N.A. Cumpsty, Review-a critical review of turbomachinery noise. J. Fluids Eng. **99**(2), 278–293 (1977)
39. X. Nogueira, S. Khelladi, I. Colominas, L. Cueto-Felgueroso, J. París, H. Gómez, High-resolution finite volume methods on unstructured grids for turbulence and aeroacoustics. Arch. Comput. Methods Eng. **18**(3), 315 (2011). https://doi.org/10.1007/s11831-011-9062-9. (ISSN:1886-1784)
40. H.-B. Li, Z.-L. Sun, X. Peng, Turbocharger noise prediction using broadband noise source model. J. Beijing Inst. Technol. (Engl. Ed.) **19**(3), 312–317 (2010)
41. STAR-CCM+. Release version 7.02.011. CD-adapco (2012), http://www.cd-adapco.com
42. STAR-CCM+. Release version 9.02.005. CD-adapco (2014), http://www.cd-adapco.com
43. M. Peric, Acoustics and turbulence: aerodynamics applications of STAR-CCM+, in *South East Asian Conference* (2012)
44. I. Tomita, S. Ibaraki, M. Furukawa, K. Yamada, The effect of tip leakage vortex for operating range enhancement of centrifugal compressor. J. Turbomach. **135**(5), 8 (2013). https://doi.org/10.1115/1.4007894

Chapter 2
Methodology for Experimental Validation

2.1 Introduction

This book is devoted to the CFD analysis of flow-induced acoustics of turbocharger compressors, as described in Chap. 1. Since this topic has not been covered by many researchers, experimental measurements are used to assess the ability of these simulations to capture compressor noise generation. In this way, the turbocharger test rig should be modeled and experimental probes should be replicated in CFD. However, the existence of long ducts in the rig greatly increases the computational effort [1].

In spite of simulating the whole test rig, shorter inlet and outlet ducts will be employed. The length of the model pipes is selected so that the location of the boundary conditions do not affect to the domain of interest (the compressor). It is not straightforward to compare numerical and experimental spectra that are not obtained at the same location, and a methodology to perform a meaningful comparison between such signals should be therefore developed.

In this chapter, the validation of a 3D CFD model of a centrifugal compressor with the objective of noise prediction against experimental measurements is described. In Sect. 2.2, the experimental setup and the procedure for the acoustic characterization of the turbocharger is described. The standard and *acoustic* compressor maps are obtained, followed by a discussion on experimental measurements performed by other researchers. The baseline numerical model, with a configuration similar to the one used by Mendonça et al. [2], is defined in Sect. 2.3. Section 2.4 is devoted to the development of a methodology to compare the spectra predicted by the simulations against experimental ones. Finally, chapter conclusions are commented in Sect. 2.5.

2.2 Experimental Methodology

In this section, all the information related to the experimental measurements used throughout this book is provided, starting with a literature review on experimental acoustic characterization of centrifugal compressors.

© Springer International Publishing AG 2018   13
R. Navarro García, *Predicting Flow-Induced Acoustics at Near-Stall Conditions in an Automotive Turbocharger Compressor*, Springer Theses, https://doi.org/10.1007/978-3-319-72248-1_2

2.2.1 Literature Review

Evans and Ward [3] studied turbocharger generated whoosh noise. The noise transmission path was analyzed and the compressor outlet hose was found to be the main source of radiated noise. Increasing hose thickness or mass gave significant reductions in the radiation of whoosh noise. Broadband resonators also provided similar benefits.

Teng and Homco [4] investigated radiated noise on a powertrain dyno in a semi-anechoic cell measured at full load conditions. Different countermeasures to reduce whoosh noise were evaluated on the dyno, i.e., the use of compressors with different trim or pre-whirl. The primary focus of this measures was on reducing the excitation source by improving the compressor surge margin. The effect of various resonators at different locations of the intake line on whoosh noise attenuation was also studied.

Sevginer et al. [5] conducted a experimental study in order to detect and eliminate two types of noise generated by the turbocharger, named as whoosh and blow noise. Measurements were performed in a test vehicle and in an engine test rig. A broadband resonator was successfully used to diminish whoosh noise in the compressor outlet.

Ha et al. [6] measured the internal pressure fluctuation of a centrifugal compressor at different operating conditions at three stages: before the leading edge, after the trailing edge and after the diffuser. External microphones were also used. Sound Pressure Level (SPL) of impeller inlet transducer presented a sudden increase when moving from best efficiency point (BEP) to stall conditions, due to the onset of flow detachments and backflows. Impeller and diffuser outlet spectra are alike: they show a broadband increase in 2–4 kHz frequency range when reducing mass flow from BEP. At stall conditions, a tone appears at 90% of rotation order, and is attributed to rotating stall by Ha et al. [6]. Orifice noise measured with external microphones is consistent with compressor inner probes spectra: amplitude in low frequency range is increased as mass flow is reduced.

Liśkiewicz et al. [7] conducted an experimental of centrifugal blower acoustic signature study in which pressure spectral maps were obtained at different probe locations for different operating conditions. Subsynchronous narrow band disturbances were found even at stable points. Before the onset of surge-related noise, the acoustic signature notably changes at the point at which pressure ratio does not increase when mass flow is reduced. Liśkiewicz et al. [7] attributed this change to inlet recirculation.

2.2.2 Turbocharger Noise Measurements

Section 2.2.1 has shown that various methods for the acoustical analysis of radial turbochargers are available in the literature. These measurements are performed to provide external radiated noise [3, 4], orifice noise [8, 9], and internal (in-duct) flow noise [3, 4, 10, 11]. Other experimental studies such as the ones carried out by Mongeau et al. [12] or Raitor and Neise [13] are based on a detailed instrumentation of the impeller with pressure transducers. This approach is quite challenging for passenger car turbocharger compressors due to their small size.

External and orifice noises are usually measured using free field microphones or portable spectrum analyzers [8]. For in-duct measurements, piezoelectric pressure transducers are the preferred choice.

Some authors [8, 9] rely on single sensor measurements for either external and internal measurements, while others prefer sophisticated methods involving more sensors in order to obtain spatially averaged noise on the external case [3, 4], or to allow the use of wave decomposition techniques on the internal case [10, 11].

2.2.3 Selected Methods

The present chapter is devoted to the comparison between experimental measurements and a CFD model. The latter only simulates the in-duct flow (and not chamber ambient through which sound radiation could be resolved), so that only in-duct experimental measurements will be covered in this work.

Both single sensor and multi sensor methods were considered, in order to evaluate their adequacy when comparing experimental data with numerical results. Single sensor methods were applied using one sensor from a multi sensor array.

Two linear arrays of three piezoelectric sensors each were designed in accordance with Piñero et al. [14], who presented a beamforming-based signal analysis method that allows the decomposition of the pressure signal into a forward-traveling wave $p_{forw}(t)$ and a backward-traveling wave $p_{back}(t)$.

Power Spectral Density (PSD) of these pressure components will be used in Sect. 2.4 as a means of validation for numerical predictions. However, the decomposition method relies on the plane wave assumption, so it is limited to low and medium frequencies. Moreover, the spacing of the sensors defines a Nyquist-type upper frequency limit [15] which must be considered to avoid aliasing problems. These constraints will also be covered in Sect. 2.4.

This decomposition technique is also used to compute the sound intensity spectrum I in the ducts according to the expression proposed by Holland and Davies [16]:

$$I = I_{forw} - I_{back} = \frac{1}{\rho_0 a_0} \left(|\hat{p}_{forw}|^2 (1 + M)^2 - |\hat{p}_{back}|^2 (1 - M)^2 \right) \quad (2.1)$$

where $\hat{p}_{forw}$ and $\hat{p}_{back}$ denote complex spectra of forward and backward pressure waves, ρ_0 and a_0 are the mean density and sound speed, and M represents the mean Mach number in the duct.

2.2.4 Experimental Apparatus

The experimental work conducted in this investigation was carried out at a facility of CMT—Motores Térmicos that hosts a large anechoic chamber and a Diesel-powered gas stand.

In order to properly isolate its acoustical properties, the turbocharger was installed at the center of the anechoic chamber, being powered from the outside by the nearby gas stand and fed with oil from a pump located in the adjacent utility room. The gas stand used to power the turbocharger has been previously described by Galindo et al. [17]. Its core is a 10 L Diesel engine which powers a 150 kW screw compressor.

Compressed air passes through a water intercooler and a settling tank to ensure cold and steady conditions. Then, it is split to both supercharge the engine and power the turbine of the turbocharger group in the anechoic chamber, which discharges at low speed through an insulated reservoir. The centrifugal compressor studied in this work is driven by the aforementioned turbine. The compressor takes its air from the chamber ambient and discharges to the adjacent auxiliary room. A layout of the test rig is depicted in Fig. 2.1.

For this investigation, the compressor was installed with inlet and outlet geometries consisting of long straight steel pipes, in order to have the simplest geometry for CFD modeling. A schematic of this arrangement can be seen in Fig. 2.2, in which piezoelectric sensor arrays are depicted and the domain modeled using CFD is highlighted.

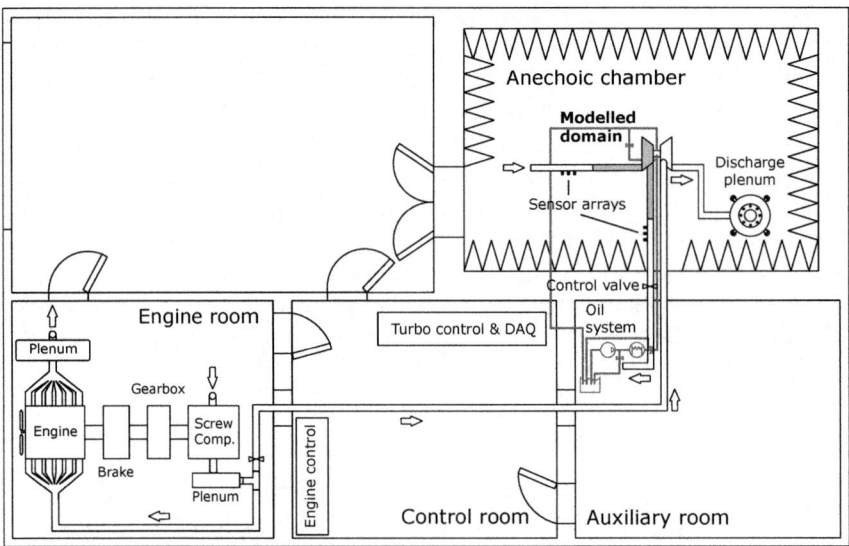

Fig. 2.1 Test rig layout (reprinted from [18])

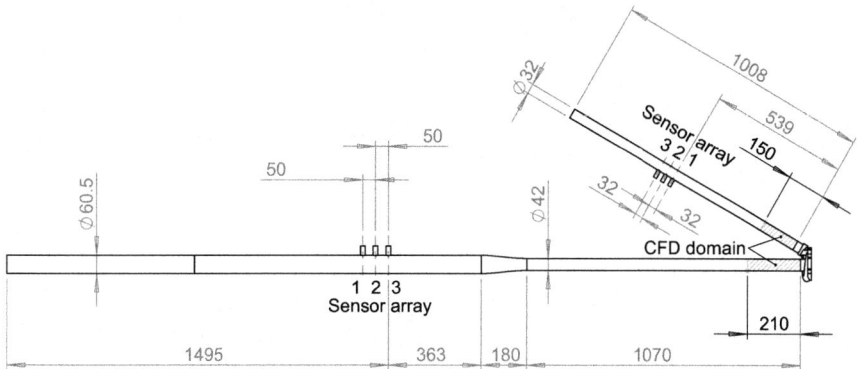

Fig. 2.2 Compressor piping sketch, highlighting the CFD domain and the sensor arrays (dimensions in mm) (reprinted from [19], with permission from Elsevier)

Two separate data acquisition systems have been used to perform the experimental measurements. In the first place, a custom-built control system monitors averaged pressures, temperatures, mass flows and engine speeds. It is also used to control the valves that set the compressor operating point, oil heating, etc. A 5 s average of every sensor is recorded for each measured data point.

Figure 2.2 also includes two arrays of piezoelectric pressure sensors which are mounted on both the inlet and outlet pipes and sampled at 100 kHz during 1 s for each data point, using the second acquisition system, a Yokogawa digital oscilloscope. These arrays enable the use of the wave decomposition technique mentioned in Sect. 2.2.3.

For a further discussion on the experimental apparatus employed for the acoustic characterization of automotive compressors, please see the works of Torregrosa et al. [20, 21] and García-Tíscar [22].

2.2.5 Compressor Map

A set of data points was measured using this setup. The procedure involved a step-by-step reduction of air mass flow, attained with the operation of the back pressure valve, while a certain turbo speed was maintained regulating the operating pressure of the screw compressor at the gas stand.

For each turbo isospeed line, data was captured between fully-opened back-pressure valve conditions and the start of deep surge. This process was repeated for 80, 100, 120, 140, 160 and 170 krpm. The number of points at the last speed is limited due to operational constraints of the facility.

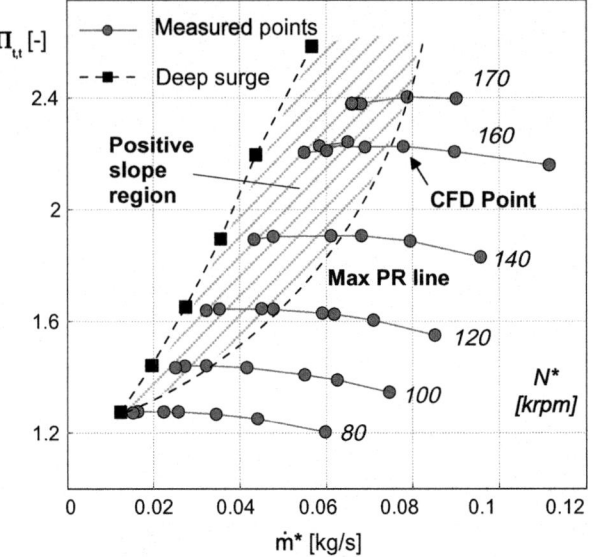

Fig. 2.3 Measured compressor map, including the simulated point and maximum compression line (after [19])

To account for the variation in ambient conditions during the measurements, air mass flow and turbocharger speed were corrected to Standard Day conditions[1]:

$$N^* = N\sqrt{\frac{T_{ref}}{T_{in,0}}} \qquad \dot{m}^* = \dot{m}\left(\frac{p_{ref}}{p_{in,0}}\right)\sqrt{\frac{T_{in,0}}{T_{ref}}}. \tag{2.2}$$

Turbocharger speed N, air mass flow $\dot{m}$, total inlet pressure $P_{in,0}$ and temperature $T_{in,0}$ were computed using averaged sensor data from the rig control system.

For each operating condition, the total-to-total pressure ratio

$$\Pi_{t,t} = \frac{p_{out,0}}{p_{in,0}} \tag{2.3}$$

was obtained from the same system, and plotted against the corrected air mass flow $\dot{m}^*$, as shown in Fig. 2.3, commonly referred to as *compressor map*.

In this kind of map, the leftmost points of each speed line conform the so-called *deep surge line*. Although this limit is usually regarded as a property of the compressor, previous experience [23–25] shows that compressor piping conditions have a great influence on surge onset.

These deep surge points for the selected compressor and inlet geometry were measured in accordance with the method developed by Galindo et al. [26] (see

---

[1] $T_{ref} = 288.15$ K and $P_{ref} = 101\,325$ Pa.

specially Fig. 6 of their paper) using the instantaneous pressure recording of the first array sensor. The measurements were carried out in a separate non-anechoic facility capable of higher power delivery to the turbine at all speeds.

Before surge line, however, there is usually a point where the slope of the speed line comes to zero or even changes its sign. The locus of points that, for a constant compressor speed, present maximum compression ratio will be named as the *maximum compression line*.

Therefore the measured compressor map is presented in Fig. 2.3 together with the maximum compression line. Also, the operation point marked in the map is the selected one for the simulation carried out using the numerical model which is described in Sect. 2.3.

2.2.6 Discussion on Experiments

The relation between whoosh noise intensity and compressor operating point is not clear, according to literature review.

Evans and Ward [3] found that whoosh noise is not encountered at the area where the operating line is closest to the surge line but at the region of the compressor map where, for a constant compressor speed, the pressure ratio across the compressor increases with increasing mass air flow (compressor speed lines present positive slope).

This zone is referred by Evans and Ward [3] as "marginal surge" or "soft surge", and it is delimited by surge line and maximum compression line. Since no actual surge exists in this zone, it will be named as *positive slope region* in this book, as shown in Fig. 2.3. Sevginer et al. [5] and Teng and Homco [4] also found whoosh noise when compressor was running at positive slope region.

Figurella et al. [8] studied the acoustic and performance characteristics of an automotive centrifugal compressor on a steady-flow turbocharger test bench. As the flow rate is reduced and the slope of the speed lines becomes less negative, the compressor exhibited a broadband elevation of noise in the 4–12 kHz band (whoosh noise), which was evident both in the upstream compressor duct and external sound pressure level (SPL) measurement locations. When the mass flow rate is decreased beyond a critical value, the temperature near the inducer tips increases sharply, suggesting local flow reversal, and the total SPL in the 4–12 kHz range suddenly reduces.

The selection of the frequency range of 4–12 kHz as whoosh noise window and the measurement of noise in the inlet duct instead of the outlet duct, in which whoosh noise is more intense [3], could be the reasons to explain the different region of whoosh noise preponderance in the work of Figurella et al. [8] compared to previous authors' work.

In order to quantify compressor hiss noise, Gaudé et al. [10] obtained a compressor map in which acoustic intensity is shown as a scalar field. Acoustic intensity was measured by applying the three microphones method and the decomposition of the acoustic measurement into an incident and reflected wave. Anechoic terminations were installed at the inlet and outlet of the compressor mounted on a turbocharger test bench. The value of acoustic intensity was averaged over the frequency range of hiss noise (500–2000 Hz). Gaudé et al. [10] observed that, for a constant compressor speed, hiss noise intensity decreases when moving away from surge line. Moreover, this noise can be higher in the inlet duct than in the outlet one in some areas of the map.

Using Eq. 2.1, sound intensity levels (SIL) are calculated for measured points. Then, SIL are averaged over 1–3 kHz window and interpolated over the whole measuring region to obtain a *sound map* similar to that of Gaudé et al. for both the inlet and outlet pipes, as pictured in Fig. 2.4.

According to Fig. 2.4, whoosh noise in the outlet duct is louder than in the inlet. In a speed line, higher intensity is found with lower mass flows. For a given mass flow rate, noise is more intense with increasing compressor speed, although this trend can be different for higher speeds (not measured due to experimental rig limitations).

Galindo et al. [27] conducted an experimental campaign on an engine test bench to determine the impact of compressor inlet geometry on its performance, surge margin and noise emission. 4 geometries were analyzed (straight duct, tapered duct, convergent nozzle and convegrent-divergent nozzle), and inlet SIL compressor maps were obtained, being all of them qualitatively similar to the one depicted in Fig. 2.4.

According to Fig. 2.4, positive slope region is the zone with highest whoosh noise. In fact, maximum compression line seems perpendicular to intensity gradient and rate of noise increase is higher in negative slope region (intensity isolines are closer in this zone). However, whoosh noise is louder in the surge line than in the maximum compression line, unlike in Evans and Ward [3] work.

There are two factors that may explain the difference. First, Fig. 2.4 shows an increase of noise with compressor speed, so operating condition with lower speeds could present less whoosh noise despite being close to surge line, which is not considered by Evans and Ward. Second, surge line shown in Evans and Ward compressor map is not a straight line, so displayed surge margin for operating points at low speeds may not be realistic because actual surge line could be shifted towards lower mass flows at these conditions. For instance, Gaudé et al. [10] performed steady measurements at the left of the surge line corresponding to the standard compressor map.

Compressor manufacturers define the surge line with a particular piping configuration (commonly straight inlet and outlet ducts). However, actual inlet hose can have an impact on compressor surge line [24, 25, 27]. Moreover, centrifugal compressor surge line is shifted towards lower mass flows when it operates on the engine with respect to flowcharts obtained using steady gas stand tests, due to pulsating flow effects [26]. Therefore, there is an uncertainty when estimating the actual surge margin of an operating condition.

In order to investigate whoosh noise generation, a detached eddy simulation is performed in the point corresponding with highest pressure ratio at 160 krpm, marked in Fig. 2.3. In the following sections, the CFD model noise prediction capabilities will be assessed by the experimental measurements already presented.

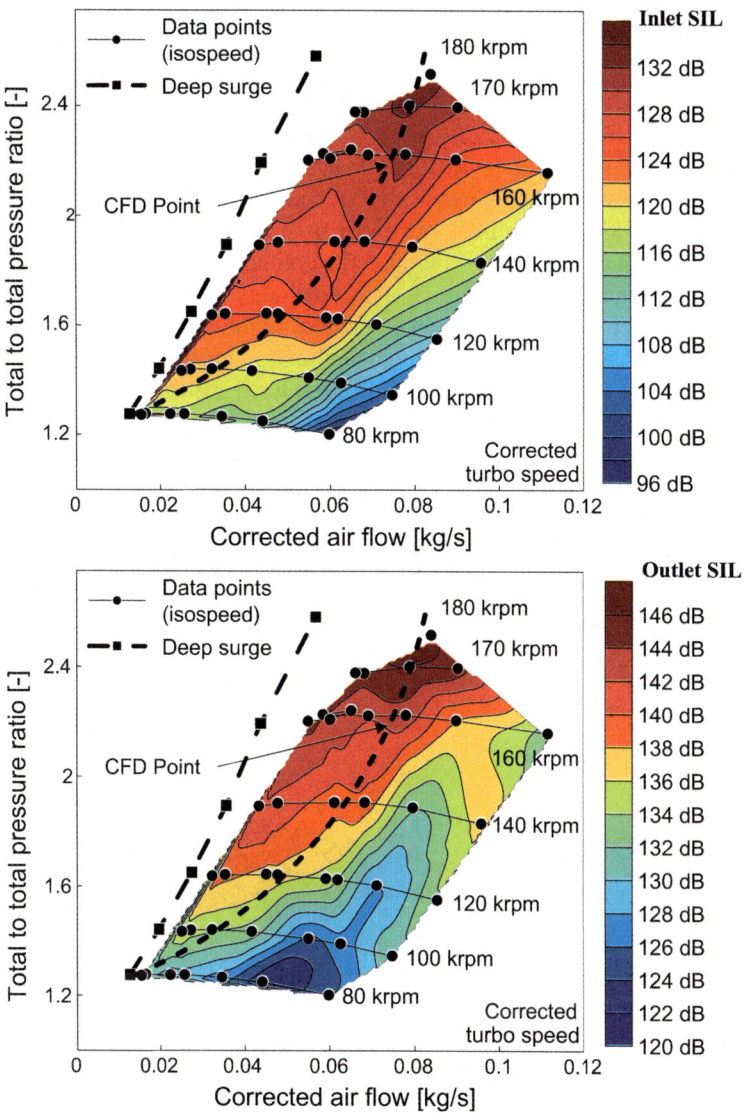

Fig. 2.4 SIL maps for inlet (top) and outlet (bottom) pipes at the selected frequency range of 1–3 kHz, including the simulated point and maximum compression line (after [18])

2.3 Numerical Model

A numerical model of the compressor was built using Star-CCM+ [28]. The impeller, vaneless diffuser and volute of the compressor were digitalized. A structured-light 3D scanner provided a point cloud of the bodies, which were processed by reverse engineering software in order to reconstruct smooth surfaces. In order to avoid influence of manufacturing variability, only one main blade and one splitter blade were digitalized, creating the rest by revolution. The compressor wheel consists of 6 full blades and 6 splitter blades. The tip clearance and the narrow gap between the rear part of the impeller and the backplate are included in the model.

Straight inlet and outlet ducts were created extruding the corresponding cross section 5 diameters long. These ducts are considered instead of the actual ones forming the turbocharger test rig (see Fig. 2.2) because it is estimated that the computational effort would have increased about 10 times if the whole rig had been simulated. The modeled domain is depicted in Fig. 2.5.

The numerical configuration is based on the previous work by Mendonça et al. [2]. The mesh consists of 9.5 million polyhedral cells. Figure 2.6 shows the rotor mesh, which was built so as to obtain y^+ values close to the unity at the impeller. Blade tip clearance along with backplate region are considered in the model, as depicted by Fig. 2.6.

The segregated solver was used to perform a detached-eddy simulation with a SST $k - \omega$ turbulence model, "which functions as a sub-grid-scale model in regions

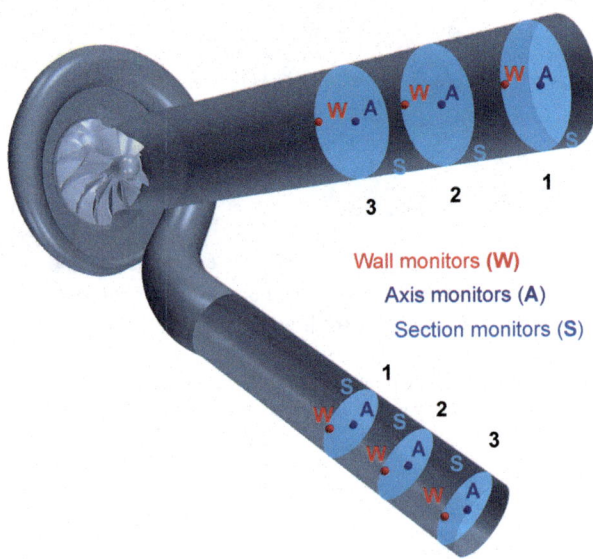

Fig. 2.5 CFD modeled domain, showing the two arrays of virtual probes located at the inlet and outlet ducts (reprinted from [19], with permission from Elsevier)

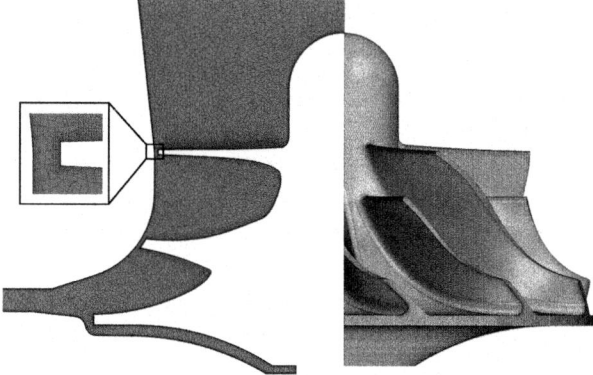

Fig. 2.6 Rotor region mesh: cross-section (left) and impeller (right), showing a detail view of tip clearance and the boundary layer mesh inflation (reprinted from [19], with permission from Elsevier)

where the grid density is fine enough for a large-eddy simulation, and as a Reynolds-averaged model in regions where it is not" [29]. Particularly, IDDES [30] is used, which combines WMLES and DDES hybrid RANS-LES approaches. The pressure was set at the outlet as a boundary condition and, unlike Mendonça et al. [2], a mass flow boundary condition was used at the inlet.

An unsteady, rigid body motion simulation was carried out to model the working point marked in Fig. 2.3. Total temperature and mass flow rate are set at the inlet, the latter being 77 g/s, which is 1.8 times the one corresponding to surge at this compressor speed (also imposed in the simulation). A pressure boundary condition is used at the outlet.

Second-order accurate transient solver employs a time-step size of 1.046×10^{-6} s, so that the impeller mesh turns $1°$ per time step at the selected operating condition. Heat transfer with the surroundings is neglected by considering adiabatic walls. Serrano et al. [31] showed that heat transfer is only a relevant fraction of the power absorbed by the compressor for low to medium loads. Therefore, prediction of compressor outlet temperature [32] or measurement of isentropic efficiency [33] are not influenced by heat transfer for high compressor speeds, such as the one studied in this work. The effect of roughness is neglected as well, i.e., smooth walls are considered in the CFD simulations.

2.4   Validation Methodology

2.4.1   Global Variables

In order to validate the numerical model described in Sect. 2.3, the results obtained by the model are compared with the experimental measurements corresponding to the simulated operating condition. Specific work and isentropic efficiency

$$W_u = \frac{\dot{W}}{\dot{m}} = \frac{\frac{2\pi N(rpm)}{60}\tau}{\dot{m}} = c_p(T_{out,0} - T_{in,0})$$

$$\eta_s = \frac{\dot{W}_{is}}{\dot{W}} = \frac{T_{in,0}\left(\Pi_{t,t}^{\frac{\gamma-1}{\gamma}} - 1\right)}{T_{out,0} - T_{in,0}} \tag{2.4}$$

are calculated in addition to compressor map variables (Eqs. 2.2–2.3) to assess the ability of the model to predict the overall behavior of the turbomachine. These variables are time averaged in both experimental measurements and transient simulations.

The comparison between measured and predicted global variables is made in Table 2.1, in which relative difference for a generic variable ϕ is defined as

$$\epsilon_{exp}(\%) = \frac{\phi_{CFD} - \phi_{exp}}{\phi_{exp}} \cdot 100. \tag{2.5}$$

Since outlet pressure is imposed in the simulation, the calculation of inlet stagnation pressure determines the error in prediction of pressure ratio. Similarly, torque (or outlet stagnation temperature) computation dictates specific work estimation and efficiency is affected by both inlet total pressure and outlet stagnation temperature.

Although the agreement in terms of global variables is excellent (relative errors do not exceed 1% in Table 2.1), further validation is required to accept the results provided by the CFD model as a means to study the noise generated by the flow field patterns inside the compressor. For instance, Hemidi et al. [34, 35] showed that, for a supersonic air ejector, validation based on global variables does not guarantee proper prediction of local flow features.

Flow measurements inside the device are common for large turbomachinery (e.g., Ubaldi et al. [36]), but for automotive turbochargers it is quite difficult to place sensors due to the small size of the device.

Table 2.1 Relative difference between experimental and numerical compressor global variables for 77 g/s case

	$\Pi_{t,t}$	W_u	η_s
$\epsilon_{exp}(\%)$	−0.9	−0.8	−0.4

After [19]

Hellström et al. [37] performed PIV measurements of the flow upstream the inducer of a centrifugal compressor with ported shroud at near surge conditions, being able to capture the main flow structures at this location with a large eddy simulation. In a continuation of this work, Semlitsch et al. [38] confirmed LES potential by showing good quantitative agreement with PIV velocity data at different cross-sections upstream of the impeller, for both stable and near-surge operating conditions. Even though recent studies have been able to obtain PIV-based aeroacoustic predictions, mostly for external flow applications [39], a direct comparison between numerical and experimental unsteady pressure measurements is preferred in this work.

Choi et al. [40] obtained experimental pressure traces with miniaturized Kulite sensors to validate a simulation of rotating stall in a transonic fan, placing the pressure probes close to the fan blades. However (instrumentation difficulties aside) Choi et al. compared rotational speed of stall cells whereas this work is interested in the evaluation of whoosh noise prediction, which is a problem for car manufactures due to excitation and subsequent radiation of inlet and outlet hoses [3].

Therefore, the Power Spectral Density of the pressure signals at both inlet and outlet ducts will be used as the tool to assess the quality of the noise prediction delivered by the model.

2.4.2 Pressure Spectra

As explained in Sect. 2.2.4, an array of three piezoelectric sensors was placed at both the inlet and the outlet ducts of the actual turbocharger. Unfortunately, the computational cost of simulating the whole system is prohibitive, and the array of three virtual probes was placed at the extruded ducts of the model, following the recommendations of Piñero et al. [14] to select their axial location. Three different monitors were investigated: two point monitors located respectively at the duct axis and next to the wall and a cross-section monitor. A sketch of the monitors is depicted in Fig. 2.5.

PSD can be computed from the raw pressure signal registered by a piezoelectric sensor/virtual monitor. Experimental signals were recorded during 1 s, whereas numerical monitors were stored for over 60 ms (corresponding to 160 impeller revolutions) after reaching a steady state in terms of global variables. Welch's overlapped segmented average [41] is used to estimate the PSD. Blocks with 50% overlap are tapered using Hamming function, although no differences were found between most common windowing functions. The number of blocks is selected so as to obtain the frequency resolution closest to 150 Hz.

In Fig. 2.7, the PSD at the inlet and outlet ducts is presented for the first probe of each array. Experimental spectrum at inlet shows an initial decay until 5 kHz and a broadband elevation from 5 to 12 kHz. From 12 kHz onwards, the PSD decreases. The most relevant features for the experimental spectrum at the outlet duct are a

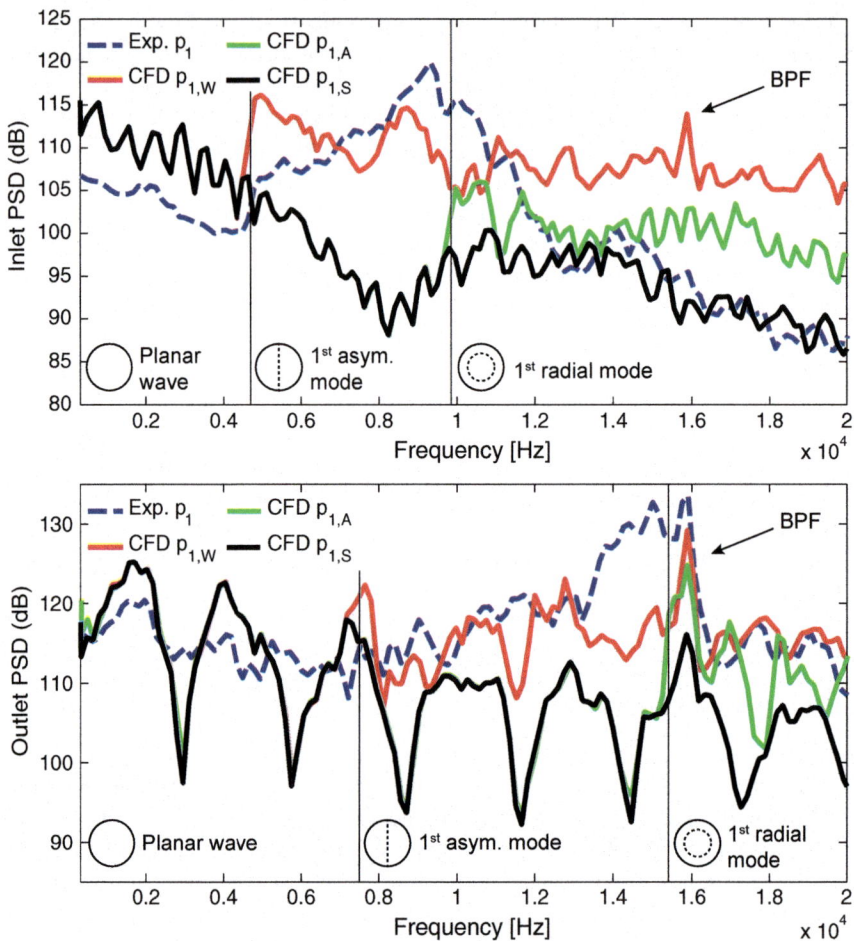

Fig. 2.7 PSD of experimental and numerical probes at inlet (top) and outlet (bottom) ducts, including the appearance of the first asymmetric and the first circular symmetric modes (reprinted from [19], with permission from Elsevier)

broadband elevation from 13 kHz until 16 kHz and the tone corresponding to the main BPF (about 16 kHz for this compressor speed (160 krpm), taking into account that the impeller presents 6 main blades).

In the CFD calculation, the three types of monitor are investigated. At the inlet, all numerical monitors at same axial position provide identical results up to 4.4 kHz, whereas the threshold increases to 7 kHz for the outlet. Similarly, axis and cross-section spectra are coincident until 9.5 kHz at the inlet pipe and 15.3 kHz at the outlet.

According to Eriksson [42], first asymmetric mode and first radial mode start propagating at cut-off frequencies of:

$$f_{c,a} = 1.84 \frac{a}{\pi D}(1 - M^2)^{1/2} \qquad f_{c,r} = 3.83 \frac{a}{\pi D}(1 - M^2)^{1/2}. \qquad (2.6)$$

These expression predicts values of 4.7 and 7.4 kHz for inlet and outlet ducts, respectively, for the asymmetric mode. The radial mode starts propagating at 9.7 kHz at the inlet duct and at 15.4 kHz at the outlet. The behavior of the different types of monitor in Fig. 2.7 is thus caused by the onset of higher order acoustic modes.

The agreement between experimental and numerical spectra depicted in Fig. 2.7 is not good. The decay of the inlet spectrum until 5 kHz and the BPF tone are features captured by the CFD simulation, but numerical spectra present ripples (see outlet spectrum for cross-section monitor in Fig. 2.7) that are not present in the measurements. Moreover, if the PSD of same type of monitor at different axial positions were plotted (not shown here), these ripples would exist at other frequencies, which can be attributed to standing waves acting at certain frequencies.

Standing waves are relevant in the numerical model because the length of straight ducts tunes frequencies of 1 kHz and above. If the whole turbocharger rig depicted in Fig. 2.2 had been simulated, standing waves would have been removed from the frequency range of interest. However, simulating such long ducts would increase the overall computational effort, as stated in Sect. 2.1.

Non-reflecting boundary conditions (NRBC) could have alleviated this issue, but they are not used in this work. NRBC represent the behavior of an infinite duct, which is not exactly the case of the experimental rig, but it is more realistic than fixing the pressure at a short length after the compressor, as done in the present simulation. In this way, NRBC should erase the spurious tuning effect on certain frequencies that presents the selected duct length.

Mendonça et al. [2] used a NRBC at the inlet while a pressure boundary condition is applied at the outlet. Torregrosa et al. [43] and Galindo et al. [1] showed that Riemann-based NRBCs, such as the ones existing in StarCCM+ or Ansys-Fluent, do not reproduce faithfully the behavior of an infinite duct for non-homentropic flows. In any case, NRBC were not used in this work for different reasons. When a NRBC was set at the outlet, the operating condition instantaneously changed, eventually reaching compressor surge. Creating a longer outlet duct and considering radial equilibrium in the outlet NRBC [1] may increase the stability of the simulation with this kind of BC. The case with inlet NRBC had to be first run using mass flow inlet BC so as to match the experimental operating point. It eventually arrives to a steady state, but the initialization process increases the computational effort. The use of inlet NRBC did not have significant impact on outlet duct spectra, but inlet spectra were not similar to those predicted by the case featuring mass flow inlet BC.

One way to overcome the problems derived from avoiding NRBC is to decompose the signals. As already said in Sect. 1.2, pressure wave decomposition is a useful tool to analyze gas exchange processes (see, for instance, the works of Torregrosa et al. [44, 45] and Serrano et al. [32, 46]). The relevant pressure components to be analyzed in this work are the ones going out of the compressor wheel (backward pressure at inlet duct and forward pressure at outlet duct). The solution is not exactly the same as with NRBC, since pressure waves are still reflected at the BC, but decomposed signals only include the information going from the compressor to the domain boundaries. Raw pressure spectra of mass flow inlet BC and case with inlet NRBC presented great discrepancies, whereas only slight differences could be noted when pressure components were used instead.

2.4.3   Spectra of Pressure Components

Figure 2.8 presents the PSD of experimental pressure signals. The decomposed pressure is obtained using the beamforming technique explained in Sect. 2.2.4. The counterpart of the decomposition is the reduction of the frequency range that can be analyzed.

A Nyquist-type criterion based on the spatial resolution of the sensor array is used to exclude aliased high frequencies. The cut-off frequency must be less than half the intrinsic frequency of the array, which corresponds to the wavelength being the separation between sensors [47]:

$$f_c < \frac{f_i}{2} = \frac{a}{2L} \qquad (2.7)$$

This limit is more restrictive than plane wave cut-off predicted by Eq. 2.6, being around 3.4 kHz for the inlet and 6.4 kHz for the outlet, considering their average sound speed and sensor separation. In any case, maximum human hearing sensitivity is comprised in this range.

PSD of three experimental probes in each duct shown at Fig. 2.8 are quite similar. Decomposed pressure PSD presents a reduction in amplitude regarding raw signals, but no different features are observed. The high coherence between signals is a result of each array of piezoelectric sensor being placed far from the compressor: 6 diameters before and 10 diameters after any cross-section change so that the flow is fully developed. Also, the great length of the ducts avoids any standing wave at frequencies of interest.

In the numerical simulations, the beamforming technique is replicated using different types of probes arranged in a 3-array fashion (Fig. 2.5). Furthermore, the information of the flow field can be used to obtain the pressure components by means of the Method of Characteristics (MoC), as described by Torregrosa et al. [43]:

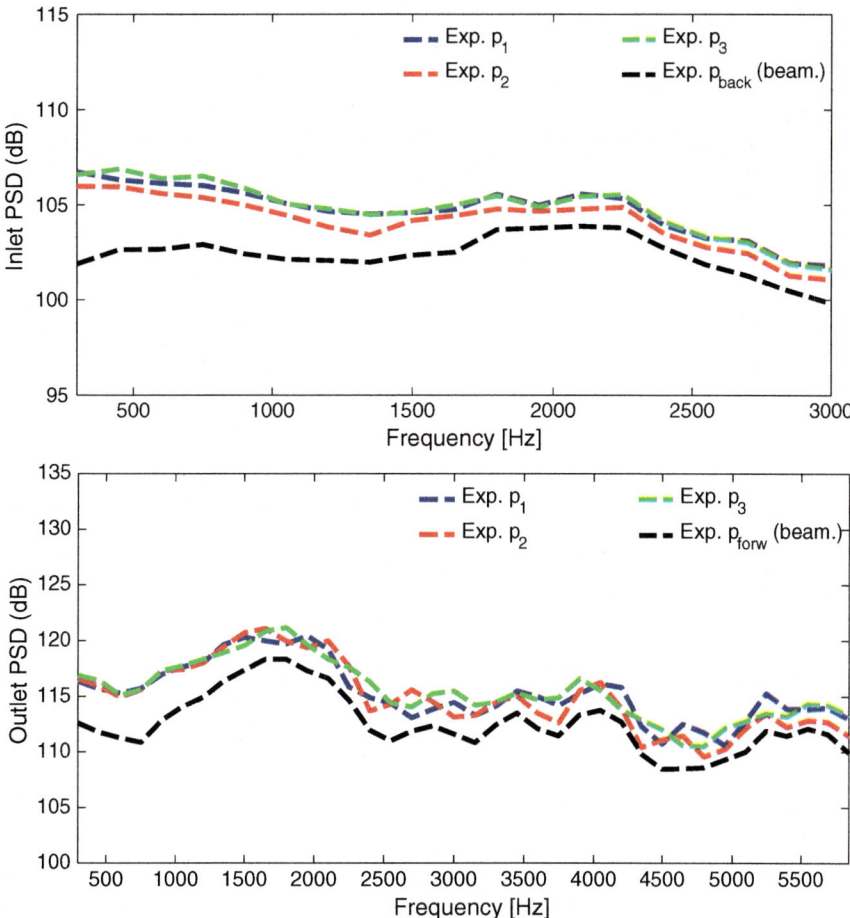

Fig. 2.8 PSD of experimental signals at inlet (top) and outlet (bottom) ducts (reprinted from [19], with permission from Elsevier)

$$p_{forw} = p_{ref} \left[\frac{1}{2} \left(1 + \left(\frac{p}{p_{ref}} \right)^{\frac{\gamma-1}{2\gamma}} \left(1 + \frac{\gamma-1}{2} \frac{u}{a} \right) \right) \right]^{\frac{2\gamma}{\gamma-1}}$$

$$p_{back} = p_{ref} \left[\frac{1}{2} \left(1 + \left(\frac{p}{p_{ref}} \right)^{\frac{\gamma-1}{2\gamma}} \left(1 - \frac{\gamma-1}{2} \frac{u}{a} \right) \right) \right]^{\frac{2\gamma}{\gamma-1}} \qquad (2.8)$$

In Figs. 2.9, 2.10 and 2.11, PSD of the signals obtained using each type of probe in the model are shown. Spectra of pressure components obtained with wall point monitors based on MoC decomposition are slightly dependent on the axial position, both

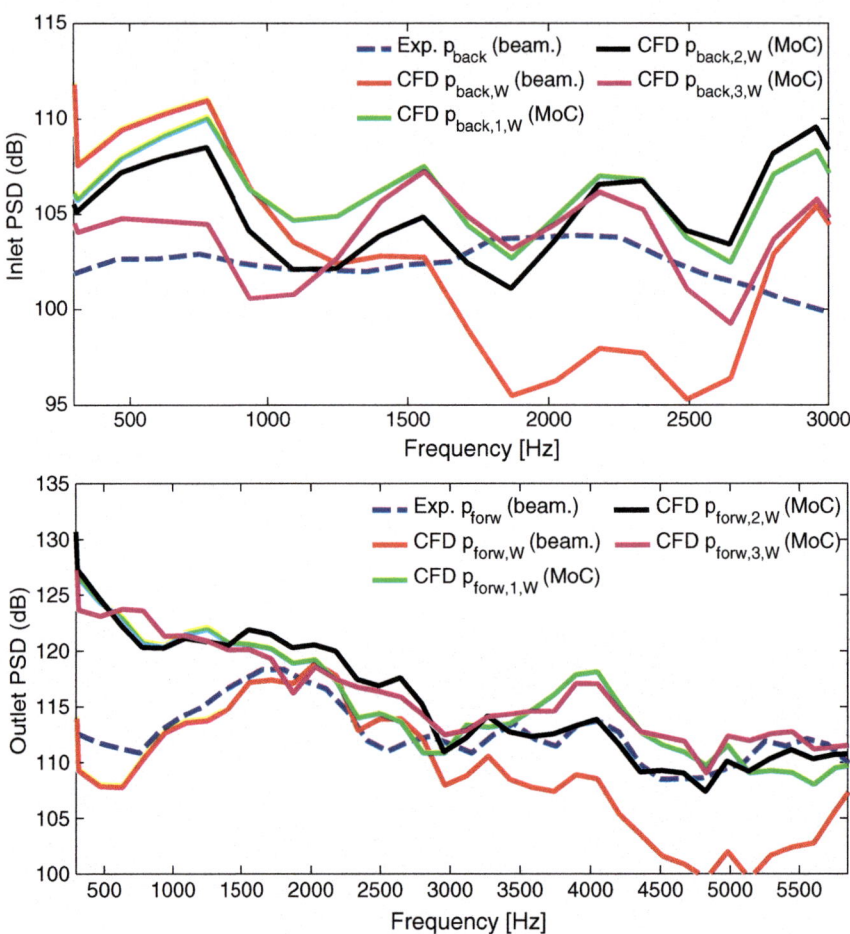

Fig. 2.9 PSD of experimental and numerical probes (wall monitors) at inlet (top) and outlet (bottom) ducts (reprinted from [19], with permission from Elsevier)

in the inlet and the outlet duct (Fig. 2.9). Moreover, PSD obtained with beamforming is different from those obtained using MoC.

Figure 2.10 shows that axis monitors provide similar PSD at the inlet duct, regardless of decomposition approach or axial position. Conversely, for these virtual sensors, forward pressure spectrum at outlet duct depends on the technique used (beamforming or MoC), and MoC based spectra are not coincident.

Inlet spectra obtained with cross-section monitors (Fig. 2.11) and axis monitors (Fig. 2.10) are alike: PSD does not depend on probe position when decomposing with Eq. 2.8 and beamforming spectrum is similar to those obtained with MoC.

In the outlet duct, PSD based on MoC is the same for all cross-section monitors, presenting features close to the ones observed with beamforming technique

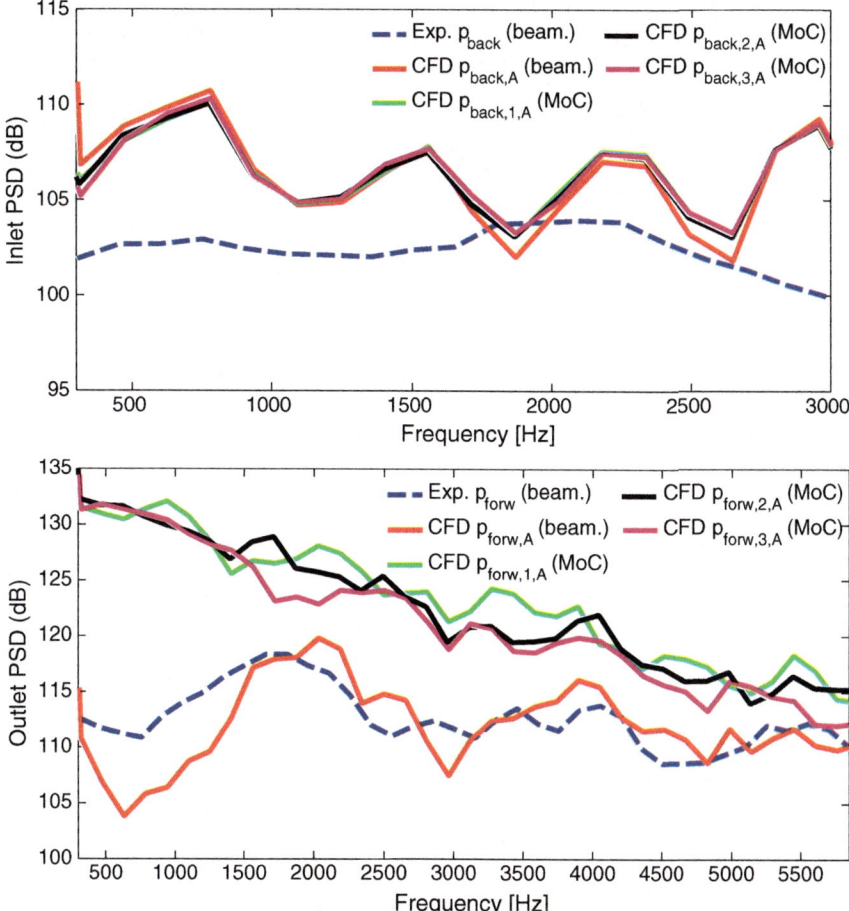

Fig. 2.10 PSD of experimental and numerical probes (axis monitors) at inlet (top) and outlet (bottom) ducts (reprinted from [19], with permission from Elsevier)

applied on axis monitors (Fig. 2.10). However, outlet forward pressure obtained using beamforming technique with cross-section monitors (Fig. 2.11) provides a spurious spectrum.

Chapters 5 and 6 will present a detailed view on the mean and transient compressor flow field, but it is useful now to anticipate some features of the flow behavior. For this near-stall operating condition, backflows appear at the inducer plane, acting as an acoustic source for the intake duct. In any case, the flow field in the vicinity of the inlet array is uniform, so that both decomposition techniques provide the same spectra (see top of Fig. 2.11). In contrast, the flow at the discharge duct is not fully developed and presents a swirling pattern, so the requirements for a proper beamforming are not

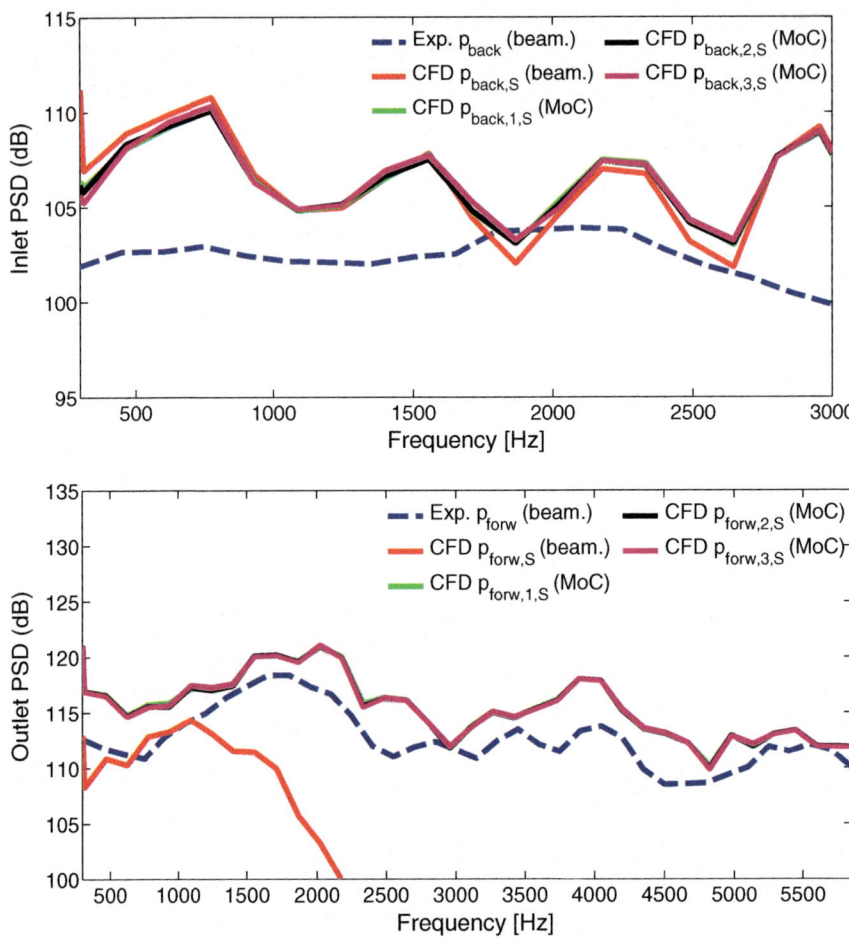

Fig. 2.11 PSD of experimental and numerical probes (cross-section monitors) at inlet (top) and outlet (bottom) ducts (reprinted from [19], with permission from Elsevier)

met at the outlet array, explaining the poor performance of this technique in bottom of Fig. 2.11.

In fact, the outlet spectra obtained with beamforming depend on the distance between the same type of virtual sensors. This is not a problem for the experimental test rig, since the arrays are placed where the flow is guaranteed to be uniform. Besides, the LCMV weights of the beamforming method are computed so as to have a strong directivity in the axial coordinate, assuming that the signals correspond to aligned points in this direction. When cross-section averages are used, the alignment is not well preserved and beamforming method may provide spurious results (see bottom of Fig. 2.11).

Regarding MoC decomposition, the irregular flow in the outlet provides a non-plane forward pressure profile, and thus wall (Fig. 2.9) and axis probes (Fig. 2.10) do not provide the same outlet spectra as when the whole cross-section is considered (Fig. 2.11).

Therefore, the selected comparison between experimental and numerical PSD of pressure component is made using beamforming for obtaining pressure components in the experimental case and Eq. 2.8 (MoC) at cross section for the numerical simulations, which are included in Fig. 2.11. At the inlet duct (upper part of the figure), experimental PSD is completely flat while numerical spectrum presents bumps centered at 750 Hz and harmonics. Using a paired difference test, a significant mean difference of 4 dB is found together with a standard deviation of 2.5 dB.

In the outlet duct (lower part of Fig. 2.11), the PSD are about one order of magnitude greater than at the inlet, in accordance with Evans and Ward [3]. A paired difference test between experimental and numerical spectra provides a mean difference of 3 dB and a standard deviation of 1.8 dB. Spectrum main features are well reproduced by the numerical model. In particular, whoosh noise is present as a broadband elevation in the experimental spectrum between 800 and 2500 Hz, in agreement with the band of 1–3 kHz used by other researchers [3, 10, 48]. Whoosh noise is also captured by the CFD simulation.

2.4.4 High Frequency Spectra

To avoid nodes of standing waves, pressure decomposition has proved to be a valuable tool in Sect. 2.4.3. Nevertheless, the employed beamforming technique sets a constraint in maximum frequency for the comparison between numerical and experimental spectra. Despite human hearing sensitivity peak lies within this range, it would be interesting to extend the validation to higher frequencies.

Observing again Fig. 2.7, it can be noted that high frequency spectra should be investigated using wall monitors, to take into account higher order modes. Raw pressure spectra could have been used because the effect of standing waves nodes above the onset of first asymmetric mode is not so important, since higher order modes prevail over plane waves. However, the impact of probe location and inlet BC (mass flow or NRBC) in pressure spectra is decreased if pressure components are considered rather than raw pressure in the simulations. In this way, PSD of virtual wall probes can be assessed as a means of validation for frequencies higher than cut-off of asymmetric mode dictated by Eq. 2.6.

Figure 2.12 shows pressure PSD obtained with each experimental probe. At the inlet, no significant differences are found between spectra of each sensor. Conversely, PSD of the third probe of the array placed at the outlet duct does not show the broadband elevation from 13 to 16 kHz that the other two sensors provide. Taking into account that spectra obtained with the outlet array does only differ in this high-frequency broadband elevation (see Figs. 2.8 and 2.12), this third sensor might have not been wall-flush mounted during the measurements, providing a different acoustic

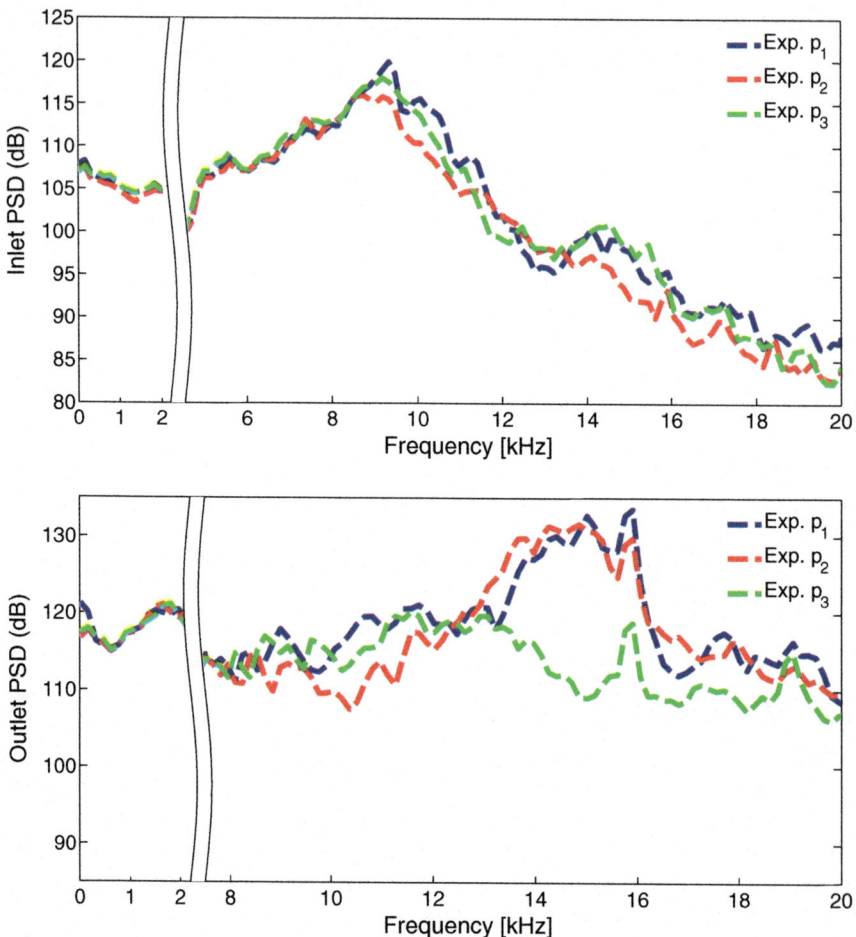

Fig. 2.12 High frequency PSD of experimental probes at inlet (top) and outlet (bottom) ducts (reprinted from [19], with permission from Elsevier)

signature for frequencies above the onset of first asymmetric mode. Hence, experimental spectrum at the outlet duct will be calculated using one of the other two probes, indistinctly.

Figure 2.13 presents spectra of MoC pressure components (see Eq. 2.8) obtained with virtual wall sensors, along with experimental spectra of first probe. PSD at inlet and outlet ducts is independent of the axial position of the monitors. Moreover, virtual wall sensors at different azimuthal positions provide spectra with no significant differences. Thus, wall monitors are a suitable choice to analyze compressor noise generation at frequencies higher than plane wave range.

Numerical inlet and outlet PSD shown at Fig. 2.13 are similar: an almost constant broadband noise with the only relevant feature of the BPF tone. The inlet broadband

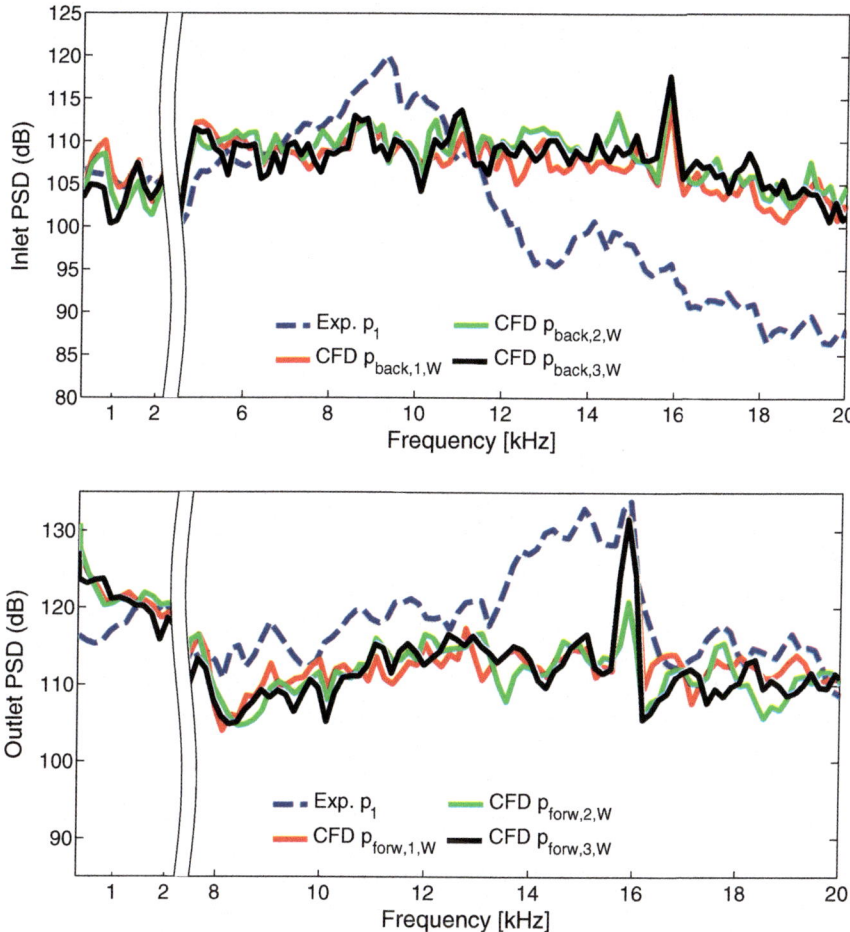

Fig. 2.13 High frequency PSD of experimental and numerical wall probes at inlet (top) and outlet (bottom) ducts (reprinted from [19], with permission from Elsevier)

elevation from 5 to 12 kHz and subsequent decay observed in the experimental spectrum is not found at numerical PSD. Conversely, the increase in amplitude between 8 and 13 kHz and the constant broadband noise above 16 kHz of the outlet experimental spectrum is accurately predicted by the simulation. The only feature of the outlet spectrum that the model is not able to reproduce is the broadband elevation from 13 to 16 kHz.

Throughout all these analysis, outlet numerical spectra are more in agreement with experimental measurements than their inlet counterparts. This could be explained by Fig. 2.2. Experimental outlet sensor array is placed about 0.4 m after the domain outlet boundary, whereas this distance extends to 1.4 m in the inlet duct, including a cross-section change. For high frequencies, mesh spacing should also be considered to explain why outlet spectrum is better predicted. Assuming that at least 20 cells are required to resolve a certain acoustical wavelength [49], mesh density at the outlet duct allows pressure waves with frequencies up to 25 kHz to propagate without numerical damping. At the inlet, this frequency reduces to 12 kHz, although BPF tone in Fig. 2.13 is satisfactorily registered. An example of the effect of numerical damping can be found in the work of Mendonça et al. [2].

2.5 Conclusions

In this chapter, the experimental measurements that are used throughout this book have been presented. Inlet and outlet ducts are instrumented with a linear array of three piezoelectric sensors. In this way, pressure spectra can be calculated using either decomposed or raw signals.

Compressor *acoustic* map shows that whoosh noise is increased as mass flow rate is reduced and compressor speed is raised. In a isospeed line, whoosh noise growth is much steeper where pressure ratio decreases with mass flow. In any case, whoosh noise is more intense at the outlet duct.

The numerical model of a centrifugal compressor inspired by the configuration of Mendonça et al. [2] has also been described. A detached eddy simulation of an operating condition at peak pressure point predicts global variables (pressure ratio, specific work and isentropic efficiency) with a relative difference less than 1% in comparison with experimental measurements.

PSD of raw pressure signals obtained with wall, axis and cross-section monitors plotted along maximum human hearing range (20–20 kHz) are coincident only in the plane-wave frequency range. Onset of asymmetric and radial acoustic modes modify the spectra according to the type of monitor.

Existence of standing waves in the simulation implies that spectra depend also on monitor axial position. Standing waves appear because the modeled domain only includes a fraction of the turbocharger test rig ducts and NRBC are not used because they either require an increase of computational effort or even lead to stability problems.

Pressure components going from the compressor to the boundaries are obtained to mitigate standing waves existing at the CFD model. In the plane wave range, decomposition of pressure signals using the Method of Characteristics with fluid variables calculated at cross-section monitors provides the most consistent signals at the simulation. The arrays of pressure sensors of the steady flow rig are placed so as to guarantee that the flow is fully developed at their position, allowing the use of beamforming for pressure decomposition of experimental signals.

Paired difference test of numerical and experimental PSD calculated in this way shows a significant mean difference of 4 dB together with a standard deviation of 2.5 dB at the inlet, whereas the differences are lower for outlet duct spectra.

In the outlet duct, the PSD is about one order of magnitude greater than at the inlet. Spectrum main features are well reproduced by the numerical model. In particular, the broadband noise existing in the experimental spectrum between 1 and 3 kHz is the so-called "whoosh noise", and is captured by the CFD simulation.

For frequencies above plane wave range, virtual wall monitors are used to capture higher order modes. Numerical spectra of pressure decomposed with MoC fail to predict a decay in amplitude for frequencies above 12 kHz at the inlet duct and a broadband noise between 13 and 16 kHz at the outlet, showing good agreement with experimental ones elsewhere.

Since the CFD simulation is able to detect whoosh noise and global variables along with pressure spectra are in agreement with experimental measurements, the current model is a good tool to predict compressor noise generation, according to the validation methodology presented in this section.

However, the sensitivity of noise prediction to model parameters should be assessed. To begin with, the current model employs the geometry as obtained by reverse engineering, but compressor working conditions (high pressure, temperature and shaft speed) modify the geometry, particularly for the case of the tip clearance. It is thus required to analyze the influence of actual tip clearance on compressor noise generation.

References

1. J. Galindo, A. Tiseira, P. Fajardo, R. Navarro, Analysis of the influence of different real flow effects on computational fluid dynamics boundary conditions based on the method of characteristics. Math. Comput. Model. **57**(7–8), 1957 (2013). https://doi.org/10.1016/j.mcm.2012.01.016. (Public Key Services and Infrastructures EUROPKI-2010-Mathematical Modelling in Engineering & Human Behaviour 2011, pp. 1957–1964. ISSN:0895-7177)
2. F. Mendonça, O. Baris, G. Capon, Simulation of radial compressor aeroacoustics using CFD, in *Proceedings of ASME Turbo Expo 2012*. GT2012-70028. ASME. 2012, pp. 1823–1832. https://doi.org/10.1115/GT2012-70028
3. D. Evans, A. Ward, *Minimizing Turbocharger Whoosh Noise for Diesel Powertrains*. SAE Technical Paper 2005-01-2485 (2005). https://doi.org/10.4271/2005-01-2485
4. C. Teng, S. Homco, Investigation of compressor whoosh noise in automotive turbochargers. SAE Int. J. Passeng. Cars-Mech. Syst. **2**(1), 1345–1351 (2009). https://doi.org/10.4271/2009-01-2053
5. C. Sevginer, M. Arslan, N. Sonmez, S. Yilmaz, Investigation of turbocharger related whoosh and air blow noise in a diesel powertrain, in *Proceedings of the 36th International Congress and Exposition on Noise Control Engineering*, 2007, pp. 476–485. ISBN:978-1-60560-385-8
6. K.-K. Ha, T.-B. Jeong, S.-H. Kang, H.-J. Kim, K.-M. Won, C.-Y. Park, W.-Y. Jung, K.-S. Cho, Experimental investigation on aeroacoustic characteristics of a centrifugal compressor for the fuel-cell vehicle. J. Mech. Sci. Technol. **27**(11), 3287–3297 (2013). https://doi.org/10.1007/s12206-013-0851-y. (ISSN:1738-494X)
7. G. Liskiewicz, L. Horodko, M. Stickland, W. Kryłłowicz, Identification of phenomena preceding blower surge by means of pressure spectral maps. Exp. Therm. Fluid Sci. **54**, 267–278 (2014). https://doi.org/10.1016/j.expthermflusci.2014.01.002
8. N. Figurella, R. Dehner, A. Selamet, K. Tallio, K. Miazgowicz, R. Wade, Noise at the mid to high flow range of a turbocharger compressor, in *41st International Congress and Exposition on Noise Control Engineering*, Vol. 2012, No. 3 (Institute of Noise Control Engineering, 2012), pp. 786–797. ISBN:978-1-62748-560-9, http://www.ingentaconnect.com/content/ince/incecp/2012/00002012/00000003/art00015

9. Y. Lee, D. Lee, Y. So, D. Chung, *Control of Airflow Noise from Diesel Engine Turbocharger*. SAE Technical Paper 2011-01-0933 (2011). https://doi.org/10.4271/2011-01-0933
10. G. Gaudé, T. Lefèvre, R. Tanna, K. Jin, T.J.B. McKitterick, S. Armenio, Experimental and computational challenges in the quantification of turbocharger vibro-acoustic sources, in *Proceedings of the 37th International Congress and Exposition on Noise Control Engineering*, Vol. 2008, No. 3 (Institute of Noise Control Engineering, 2008), pp. 5598–5611. ISBN:978-1-60560-989-8
11. H. Tiikoja, H. Rämmal, M. Abom, H. Boden, Investigations of automotive turbocharger acoustics. SAE Int. J. Engines **4**(2), 2531–2542 (2011). https://doi.org/10.4271/2011-24-0221
12. L. Mongeau, D. Thompson, D. Mclaughlin, A method for characterizing aerodynamic sound sources in turbomachines. J. Sound Vib. **181**(3), 369–389 (1995). https://doi.org/10.1006/jsvi.1995.0146
13. T. Raitor, W. Neise, Sound generation in centrifugal compressors. J. Sound Vib. **314**, 738–756 (2008). https://doi.org/10.1016/j.jsv.2008.01.034. (ISSN:0022-460X)
14. G. Piñero, L. Vergara, J. Desantes, A. Broatch, Estimation of velocity fluctuation in internal combustion engine exhaust systems through beamforming techniques. Meas. Sci. Technol. **11**(11), 1585–1595 (2000). https://doi.org/10.1088/0957-0233/11/11/307
15. A.F. Seybert, Two-sensor methods for the measurement of sound intensity and acoustic properties in ducts. J. Acoust. Soc. Am. **83**(6), 2233–2239 (1988). https://doi.org/10.1121/1.396352
16. K. Holland, P. Davies, The measurement of sound power flux in flow ducts. J. Sound Vib. **230**(4), 915–932 (2000). https://doi.org/10.1006/jsvi.1999.2656
17. J. Galindo, J.R. Serrano, C. Guardiola, C. Cervelló, Surge limit definition in a specific test bench for the characterization of automotive turbochargers. Exp. Therm. Fluid Sci. **30**(5), 449–462 (2006). https://doi.org/10.1016/j.expthermflusci.2005.06.002
18. A. Broatch, J. Galindo, R. Navarro, J. García-Tíscar, A. Daglish, R.K. Sharma, Simulations and measurements of automotive turbocharger compressor whoosh noise. Eng. Appl. Comput. Fluid Mech. **9**(1), 12 (2015). https://doi.org/10.1080/19942060.2015.1004788
19. A. Broatch, J. Galindo, R. Navarro, J. García-Tíscar, Methodology for experimental validation of a CFD model for predicting noise generation in centrifugal compressors. Int. J. Heat Fluid Flow **50**, 134–144 (2014). https://doi.org/10.1016/j.ijheatfluidflow.2014.06.006
20. A.J. Torregrosa, A. Broatch, R. Navarro, J. García-Tíscar, Acoustic characterization of automotive turbocompressors. Int. J. Engine Res. **16**(1), 31–37 (2015). https://doi.org/10.1177/1468087414562866
21. A.J. Torregrosa, A. Broatch, X. Margot, J. García-Tíscar, Experimental methodology for turbocompressor in-duct noise evaluation based on beamforming wave decomposition. J. Sound Vib. **376**, 60–71 (2016). https://doi.org/10.1016/j.jsv.2016.04.035
22. J. García-Tíscar, *Experiments on Turbocharger Compressor Acoustics*. Ph.D. thesis. (Universitat Politècnica de València, 2017), http://hdl.handle.net/10251/79552
23. J. Galindo, J.R. Serrano, X. Margot, A. Tiseira, N. Schorn, H. Kindl, Potential of flow pre-whirl at the compressor inlet of automotive engine turbochargers to enlarge surge margin and overcome packaging limitations. Int. J. Heat Fluid Flow **28**(3), 374–387 (2007). https://doi.org/10.1016/j.ijheatfluidflow.2006.06.002
24. J. Galindo, F. Arnau, A. Tiseira, R. Lang, H. Lahjaily, T. Gimenes, *Measurement and Modeling of Compressor Surge on Engine Test Bench for Different Intake Line Configurations*. SAE Technical Paper 2011-01-0370 (2011). https://doi.org/10.4271/2011-01-0370
25. J.R. Serrano, X. Margot, A. Tiseira, L.M. García-Cuevas, Optimization of the inlet air line of an automotive turbocharger. Int. J. Engine Res. **14**(1), 92–104 (2013). https://doi.org/10.1177/1468087412449085
26. J. Galindo, H. Climent, C. Guardiola, A. Tiseira, On the effect of pulsating flow on surge margin of small centrifugal compressors for automotive engines. Exp. Therm. Fluid Sci. **33**(8), 1163–1171 (2009). https://doi.org/10.1016/j.expthermflusci.2009.07.006
27. J. Galindo, A. Tiseira, R. Navarro, D. Tarí, C.M. Meano, Effect of the inlet geometry on performance, surge margin and noise emission of an automotive turbocharger compressor. Appl. Therm. Eng. **110**, 875–882 (2017). https://doi.org/10.1016/j.applthermaleng.2016.08.099

28. STAR-CCM+. Release version 9.02.005. CD-adapco (2014), http://www.cd-adapco.com
29. Detached-eddy simulations past a circular cylinder. Flow Turbul. Combust. **63**(1–4), 293–313 (2000)
30. M.L. Shur, P.R. Spalart, M.K. Strelets, A.K. Travin, A hybrid RANS-LES approach with delayed-DES and wall-modelled LES capabilities. Int. J. Heat Fluid Flow **29**(6), 1638–1649 (2008). https://doi.org/10.1016/j.ijheatfluidflow.2008.07.001
31. J.R. Serrano, P. Olmeda, F. Arnau, M. Reyes-Belmonte, A. Lefebvre, Importance of heat transfer phenomena in small turbochargers for passenger car applications. SAE Int. J. Engines **6**(2), 716–728 (2013). https://doi.org/10.4271/2013-01-0576
32. J.R. Serrano, F.J. Arnau, R. Novella, M. Á. Reyes-Belmonte, *A Procedure to Achieve 1D Predictive Modeling of Turbochargers Under Hot and Pulsating Flow Conditions at the Turbine Inlet*. SAE Technical Paper 2014-01-1080 (2014), 13pp. https://doi.org/10.4271/2014-01-1080
33. J.R. Serrano, P. Olmeda, F.J. Arnau, A. Dombrovsky, L. Smith, Methodology to characterize heat transfer phenomena in small automotive turbochargers: experiments and modelling based analysis, in *Proceedings of ASME Turbo Expo* (2014)
34. A. Hemidi, F. Henry, S. Leclaire, J.-M. Seynhaeve, Y. Bartosiewicz, CFD analysis of a supersonic air ejector. Part I: experimental validation of single-phase and two-phase operation. Appl. Therm. Eng. **29**(8–9), 1523–1531 (2009). https://doi.org/10.1016/j.applthermaleng.2008.07.003. (ISSN:1359-4311)
35. A. Hemidi, F. Henry, S. Leclaire, J.-M. Seynhaeve, Y. Bartosiewicz, CFD analysis of a supersonic air ejector. Part II: relation between global operation and local flow features. Appl. Therm. Eng. **29**(14–15), 2990–2998 (2009). https://doi.org/10.1016/j.applthermaleng.2009.03.019. (ISSN:1359-4311)
36. M. Ubaldi, P. Zunino, G. Barigozzi, A. Cattanei, An experimental investigation of stator induced unsteadiness on centrifugal impeller outflow. J. Turbomach. **118**, 41–51 (1996). https://doi.org/10.1115/1.2836604
37. F. Hellström, E. Guillou, M. Gancedo, R. DiMicco, A. Mohamed, E. Gutmark, L. Fuchs, *Stall Development in a Ported Shroud Compressor Using PIV Measurements and Large Eddy Simulation*. Technical report SAE Technical Paper 2010-01-0184 (2010). https://doi.org/10.4271/2010-01-0184
38. B. Semlitsch, V. JyothishKumar, M. Mihaescu, L. Fuchs, E. Gutmark, M. Gancedo, *Numerical Flow Analysis of a Centrifugal Compressor with Ported and Without Ported Shroud*. SAE Technical Paper 2014-01-1655 (2014). https://doi.org/10.4271/2014-01-1655
39. B.W. van Oudheusden, PIV-based pressure measurement. Meas. Sci. Technol. **24**(10), 32pp (2013). https://doi.org/10.1088/0957-0233/21/10/105401
40. M. Choi, N.H. Smith, M. Vahdati, Validation of numerical simulation for rotating stall in a transonic fan. J. Turbomach. **135**(2), 8 (2013). https://doi.org/10.1115/1.4006641
41. P. Welch, The use of fast Fourier transform for the estimation of power spectra: a method based on time averaging over short, modified periodograms. IEEE Trans. Audio Electroacoust. **15**(2), 70–73 (1967)
42. L.J. Eriksson, Higher order mode effects in circular ducts and expansion chambers. J. Acoust. Soc. Am. **68**, 545 (1980). https://doi.org/10.1121/1.384768
43. A.J. Torregrosa, P. Fajardo, A. Gil, R. Navarro, Development of a non-reflecting boundary condition for application in 3D computational fluid dynamic codes. Eng. Appl. Comput. Fluid Mech. **6**(3), 447–460 (2012). https://doi.org/10.1080/19942060.2012.11015434
44. A.J. Torregrosa, J.R. Serrano, J. Dopazo, S. Soltani, *Experiments on Wave Transmission and Reflection by Turbochargers in Engine Operating Conditions*. SAE Technical Paper 2006-01-0022 (2006). https://doi.org/10.4271/2006-01-0022
45. A.J. Torregrosa, J. Galindo, J.R. Serrano, A. Tiseira, A procedure for the unsteady characterization of turbochargers in reciprocating internal combustion engines, in *Fluid Machinery and Fluid Mechanics* (Springer, Berlin, 2009), pp. 72–79
46. J.R. Serrano, F.J. Arnau, V. Dolz, A. Tiseira, C. Cervelló, A model of turbocharger radial turbines appropriate to be used in zeroand one-dimensional gas dynamics codes for internal combustion engines modelling. Energy Convers. Manage. **49**(12), 3729–3745 (2008). https://doi.org/10.1016/j.enconman.2008.06.031

47. M. Åbom, H. Bodén, Error analysis of two-microphone measurements in ducts with flow. J. Acoust. Soc. Am. **83**(6), 2429–2438 (1988). https://doi.org/10.1121/1.396322
48. E.P. Trochon, *A New Type of Silencers for Turbocharger Noise Control*. SAE Technical Paper, vol. 110, (6), pp. 1587–1592 (2001). https://doi.org/10.4271/2001-01-1436
49. F. Mendonça, A. Read, F. Imada, V. Girardi, *Efficient CFD Simulation Process for Aeroacoustic Driven Design*. SAE Technical Paper 2010-36-0545 (2010). https://doi.org/10.4271/2010-36-0545

Chapter 3
Influence of Tip Clearance on Flow Behavior and Noise Generation

3.1 Introduction

In Chap. 2, a CFD model of the centrifugal compressor has been described, which is strongly influenced by the paper of Mendonça et al. [3]. A methodology to compare experimental and numerical spectra has been also developed, showing that the CFD model predicts noise features in agreement with those found in the experiments, for a working point at peak pressure ratio (mass flow rate of 77 g/s at 160 krpm).

The model includes the distance between the tip of the blades and the shroud, i.e., the tip clearance, as depicted in Fig. 2.6. Several researchers [2–4] consider leakage flow from blade pressure side (PS) to suction side (SS) across tip clearance as a key point in the stability and noise production of centrifugal compressors. Nevertheless, compressor geometry is often obtained from reverse engineering or manufacturer's CAD, thus modeling a *cold* tip clearance that is not representative of actual working conditions.

So far, deformations due to temperature, rotation or pressure are neglected, and shaft motion is not considered. Including these effects will led to coupled fluid-structure simulations, which are very expensive. However, the effect of these factors should be studied to assess the validity of simulations using CAD baseline clearance. In such a sensitivity study, tip clearance ratios should be set in accordance to expected values while compressor is operating.

It should be noticed that for the sensitivity studies that are conducted in this chapter and Chap. 4, the mass flow rate is not 77 g/s but 60 g/s. The selected working point (60 g/s) has been shifted towards surge line (keeping the compressor speed of 160 krpm; see Fig. 2.3) because flow is expected to be more irregular and sensitivity to numerical configuration may be increased. Conversely, peak pressure point (77 g/s) was employed in Chap. 2 because whoosh noise is more prominent in that operating point.

In this chapter, the effect of tip clearance size on fluid flow and noise spectra of a turbocharger compressor is studied. First, a literature review is performed in Sect. 3.2. Then, Sect. 3.3 is devoted to the estimation of tip clearance reduction during

© Springer International Publishing AG 2018
R. Navarro García, *Predicting Flow-Induced Acoustics at Near-Stall Conditions in an Automotive Turbocharger Compressor*, Springer Theses,
https://doi.org/10.1007/978-3-319-72248-1_3

actual compressor operation. In Sect. 3.4, the approach to perform the clearance sensitivity analysis is explained. Tip clearance size impact on global variables and noise generation is evaluated in Sect. 3.5 by means of PSD comparison. Flow field is investigated in Sect. 3.6 in order to explain the spectra obtained by different tip clearances. Finally, Sect. 3.7 includes the main findings of the chapter.

3.2 Literature Review

Raitor and Neise [2] analyzed the sound generation mechanisms of centrifugal compressors using experimental measurements, as described in Sect. 1.2. Raitor and Neise considered that secondary flow through blade tip clearance is the source of the dominating noise for subsonic flow conditions at low compressor speeds.

In this frame, CFD allows researchers to investigate centrifugal compressor flow field and particularly tip leakage flow.

Tomita et al. [4] studied two compressors with similar map except in surge vicinity. Unsteady pressure measurements performed just upstream the impeller leading edge at low mass flow rate revealed differences between compressors: the one with the narrowest operating range showed large pressure fluctuations at subsynchronous speeds moving circumferentially at about 50–80% of compressor speed, whereas compressor with wide operating range did not exhibit this large amplitude fluctuations. Tomita et al. [4] also performed 3D CFD numerical simulations in order to analyze fluid phenomena at this low mass flow rate conditions. Compressor with less surge margin presented a tornado-type vortex between blades, blocking the incoming flow. This vortex is convected downstream and a new one is created in the adjacent blade, in a rotating stall pattern. Conversely, spiral-type tip leakage vortex breakdown occurred at all blades for compressor with higher surge margin. Some rotating instability was observed in this compressor, but not as severe as the rotating stall experienced as in the other compressor. Tomita et al. [4] concluded that tip leakage vortex breakdown could stabilize the flow structure and thus increase surge margin.

Moreover, several researchers have performed studies of tip clearance sensitivity on global variables and local flow features.

Danish et al. [5] conducted a numerical study in which overall performance and flow field were investigated for nine tip clearance levels from no gap to 16% tip clearance ratio (TCR). Numerical simulations were performed with frozen rotor technique. Two different clearance increase approaches were used: reduction of blade height while keeping same housing dimensions and increase of shroud diameter with same blade height. For operating conditions close to best efficiency point, maximum increase of TCR by decrease of blade height caused a loss of 15% in pressure ratio and 10% in efficiency. If shroud size is increased instead, pressure ratio is reduced by 13% and efficiency by 8.5%.

Jung et al. [6] performed steady simulations using the same impeller with six different tip clearance profiles. Reductions in tip clearance caused an increase in pressure ratio and efficiency. Particularly, a decrease in tip clearance at the trailing edge improved the compressor performance more than implementing the same reduction in leading edge tip clearance. A reduced tip leakage flow was found to significantly improve the diffusion process downstream the impeller.

Measurements performed by Wang et al. [7] indicated that variations of tip clearance caused small influence on stall inception. However, the reasons could not be investigated because their mixing-plane, steady simulations failed to converge at near surge conditions.

However, in these sensitivity studies, particular TCRs are not usually justified by expected values of compressor actual working conditions and effect on noise generation is not assessed. The remainder of this chapter is devoted to these objectives.

3.3 Estimation of Actual Tip Clearance

In Sect. 3.1, some effects usually not considered when simulating turbocharger compressors were introduced. For the sake of accuracy, a two-way coupled fluid-structure simulation should be performed to take into account potential aeroelastic effects, along with impeller and housing deformations caused by temperature, rotation and pressure differences. However, this approach would lead to unaffordable computational times.

Wang et al. [7] used a two-step calculation procedure to take into account impeller deformation when simulating centrifugal compressor flow, avoiding a two-way coupled fluid-structure simulation. First, the deformation of the compressor blades were simulated with finite element method (FEM). Then, the flow field was recalculated with the deformed impeller estimated with FEM. In this way, tip clearance changed from a uniform clearance ratio of 7.3% into a non-uniform profile that presents greater reduction with increasing meridional coordinate (TCR is 50% larger at the leading edge than at the trailing edge). Moreover, incident and exit blade metal angles were also deformed. Particularly, tip outlet angle at trailing edge increased 3.5°. Wang et al. [7] noted that tip leakage flow is affected by impeller deformation, over all at trailing edge.

In this chapter, sensitivity of TCR to centrifugal compressor aeroacoustics will be first investigated. Aeroelasticity is neglected because centrifugal compressor blades are not as long and slender as in their axial counterparts, which do suffer from phenomena such as flutter. Deformation due to thermal and pressure loads and centrifugal forces can be taken into account with a one-way fluid-mechanical simulation. If noise is highly dependent on TCR, the flow field should be recalculated with the actual tip clearance geometry, following the approach of Wang et al. [7].

Besides, dynamic mesh approach should be employed to consider the complex shaft motion pattern described by Nelson [8]. This would imply a great increase in computational effort compared to traditional sliding mesh technique, in which rotor region is meshed only once and it is rotated across sliding interfaces, in accordance with compressor speed. The possible effect of shaft precession on noise generation is thus neglected in this chapter. Instead, eccentricity measurements are performed in Sect. 3.3.2 to assess maximum clearance reduction due to shaft motion, so as to be considered in the subsequent sensitivity analysis.

3.3.1 Thermal and Rotational Deformation

To estimate decrease of tip clearance due to impeller deformation, mean wheel temperature (120 °C) predicted by CFD and rotational speed (160 krpm) were used as an input to a structural simulation with FEM, which computed the nodal displacement. The impeller is meshed with more than 12,000 quad/tri elements. The equations are then solved using the preconditioned conjugate gradient solved implemented in ANSYS Structural.

Pressure deformation was not considered because its contribution to impeller deformation is marginal [7]. Selection of mean value instead of temperature distribution is justified by high thermal diffusivity of aluminum, which is the material of most compressor wheels [9]. The simulation was done with characteristics of the Aluminum Alloy (density = 2770 kg/m^3, Young's Modulus = 71 GPa), providing a maximum deformation for the inducer blades of 11% of CAD tip clearance.

3.3.2 Eccentricity Due to Shaft Motion

In order to estimate the tip clearance variation due to shaft motion, the technique developed by Pastor et al. [10] is applied. It consists in an image recording methodology with a processing algorithm to estimate the shaft motion. Pastor et al. [11] showed that this technique requires a less intrusive apparatus than infrared sensor measurements, also providing direct observation of shaft motion phenomenon. Galindo et al. [12] used this methodology to improve the knowledge of the behavior of the turbochargers in critical conditions of lubrication, because, as López [13] indicated, image recording is better suited than standard inductive sensors for shaft motion measurements in these lubrication conditions.

The camera was placed in front of the turbocharger on the compressor side, positioning it as coaxial as possible to the turbocharger shaft and focused to the shaft tip. A 90° elbow was attached at the inlet of the compressor housing with a glass window to allow the proper display of the shaft through the camera. Images were taken with a PCO Pixelfly 12-bit CCD Camera with spatial resolution of 1280 × 1024 pixels. The distance between the compressor and the camera was approximately 300 mm. The resolution obtained with this layout is approximately 5 μm per pixel, which allows the observation of shaft movements.

The processing algorithm consists on differentiating specific zones of the image in order to obtain their coordinates. Two screws, placed at the impeller's eye

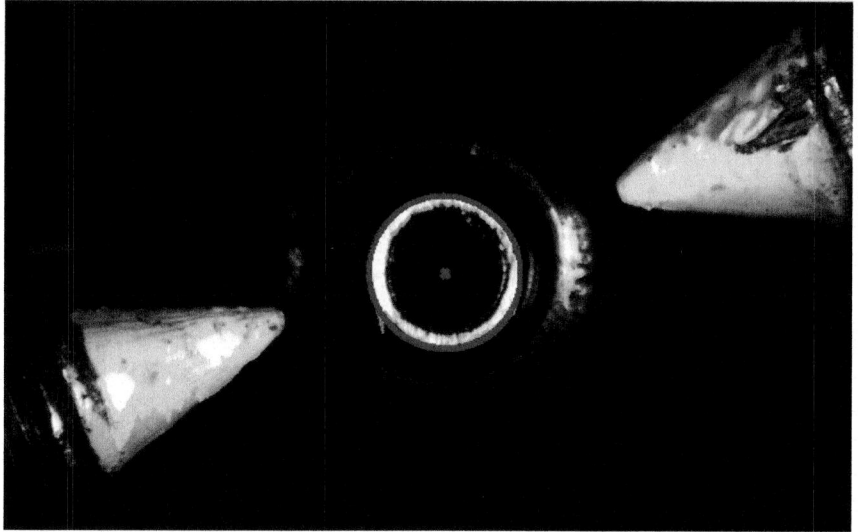

Fig. 3.1 Photograph of shaft motion, showing the estimated position of the shaft center (reprinted from [1], with permission from Elsevier)

plane, were used as reference points to calculate the relative position of the shaft, avoiding errors due to structural vibrations. Figure 3.1 shows a photograph of the shaft motion. The impeller's eye can be seen along with the reference points (left and right screws). The processing algorithm identifies the shaft tip and its center (in blue in Fig. 3.1), whose coordinates are calculated using the reference points for the subsequent determination of the shaft motion. For more details about the image-recording apparatus or the processing algorithm, please refer to the work of López [13].

For the studied compressor speed (160 krpm), the shaft departed between 100 and 150 μm regarding the center of the impeller's plane eye. This motion represents a 22–34% of reduction of the original clearance. If it is combined with the deformation calculated in Sect. 3.3.1 to set a worst-case scenario, a maximum reduction of the original tip clearance between 33 and 45% can be estimated.

3.4 Tip Clearance Reduction Approach

In Sect. 3.3, a maximum clearance reduction of 45% is postulated. In order to assess compressor performance indicators, flow features and noise production sensitivity to tip clearance, three levels are considered: original (CAD) clearance, 2/3 of original clearance and 1/3 of original clearance. In this way, expected clearance reduction range is comprised in the analysis.

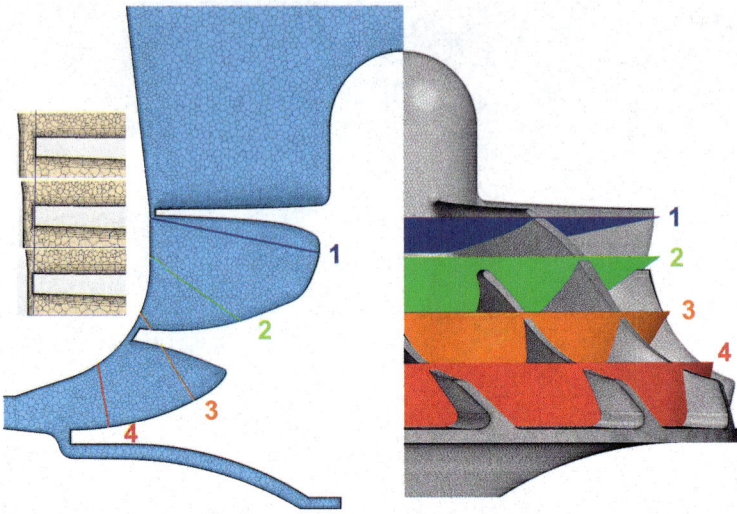

Fig. 3.2 Rotor region mesh: cross-section (left) and impeller (right), including a close-up of tip clearance for the cases studied and isomeridional surfaces (reprinted from [1], with permission from Elsevier)

Tip clearance reduction can be done with two opposed approaches, either increasing impeller size or decreasing shroud diameter. Impeller increase would mimic temperature and rotational deformation, but tip clearance reduction due to shaft motion represents an approximation regardless of the method used. However, unrealistic deformation of impeller geometry may impact flow behavior more than the associated reduction of tip clearance. In this chapter it is thus preferred to keep impeller size and reduce the shroud to obtain the desired TCRs. In any case, Danish et al. [5] proved that these two approaches to change tip clearance yield similar results.

The reduction of shroud that causes the three different TCRs studied in this section can be observed in the left-hand side of Fig. 3.2. The three meshes produced with this approach present a similar amount of elements, being about 9.5 million polyhedral cells.

3.5 Comparison of Overall Performance and Acoustic Signatures

$k - \omega$ SST URANS and DES simulations are performed with the CFD model described in Sect. 2.3, using the three TCR defined in Sect. 3.4 for each turbulence approach. Experimental measurements, obtained according to the methodology explained in Sect. 2.2, are used as a means of comparison.

Table 3.1 Experimental and numerical compressor global variables for tip clearance sensitivity analysis at 60 g/s

Case		$\Pi_{t,t}$ (−) ($\phi/\epsilon_{exp}/\epsilon_{3/3}$)	W_u (kJ · kg^{-1}) ($\phi/\epsilon_{exp}/\epsilon_{3/3}$)	η_s (%) ($\phi/\epsilon_{exp}/\epsilon_{3/3}$)
Exp.		2.22/–/–	121/–/–	62.2/–/–
URANS	3/3	2.21/−0.5/–	122/0.7/–	62.3/0.1/–
	2/3	2.26/1.7/2.2	123/1.2/0.5	63.9/2.7/2.6
	1/3	2.31/4.1/4.7	124/2.2/1.5	65.3/5.0/4.9
DES	3/3	2.23/0.4/–	122/0.6/–	63.2/1.6/–
	2/3	2.27/2.1/1.6	122/0.1/−0.5	65.0/4.4/2.8
	1/3	2.32/4.2/3.8	122/0.7/0.1	66.5/6.8/5.1

Reprinted from [1], with permission from Elsevier

3.5.1 Compressor Performance Variables

In the first place, compressor global variables are compared. Total-to-total pressure ratio, specific work and isentropic efficiency are either experimentally measured or numerically calculated using Eqs. 2.3–2.4. Table 3.1 includes these global variables along with relative difference against experimental measurements, as defined in Eq. 2.5. Moreover, relative difference against CAD clearance case with the corresponding turbulence model is also considered in Table 3.1, being defined as:

$$\epsilon_{3/3}(\%) = \frac{\phi_{CFD} - \phi_{CFD,3/3}}{\phi_{CFD,3/3}} \cdot 100. \qquad (3.1)$$

For both turbulence approaches, cases with original clearance provide compressor performance values closer to experimental results than cases with reduced clearance. Decreasing tip gap improves compressor isentropic efficiency and pressure ratio. In URANS simulations, a reduction of TCR slightly increases specific work, but in DES there is not a monotonous trend.

URANS case with CAD clearance predicts isentropic efficiency showing better agreement with experimental measurements than DES, being the rest of compressor performance parameters equivalent. However, a better estimation of global variables does not necessarily imply that URANS is more accurate than DES in predicting compressor noise generation, as the next section will prove.

3.5.2 Compressor Acoustic Spectra

In order to assess the effect of tip clearance on centrifugal compressor noise, the methodology developed in Sect. 2.4 is followed. In this way, Figs. 3.3, 3.4, 3.5 and 3.6 are obtained for the $k - \omega$ SST URANS and DES.

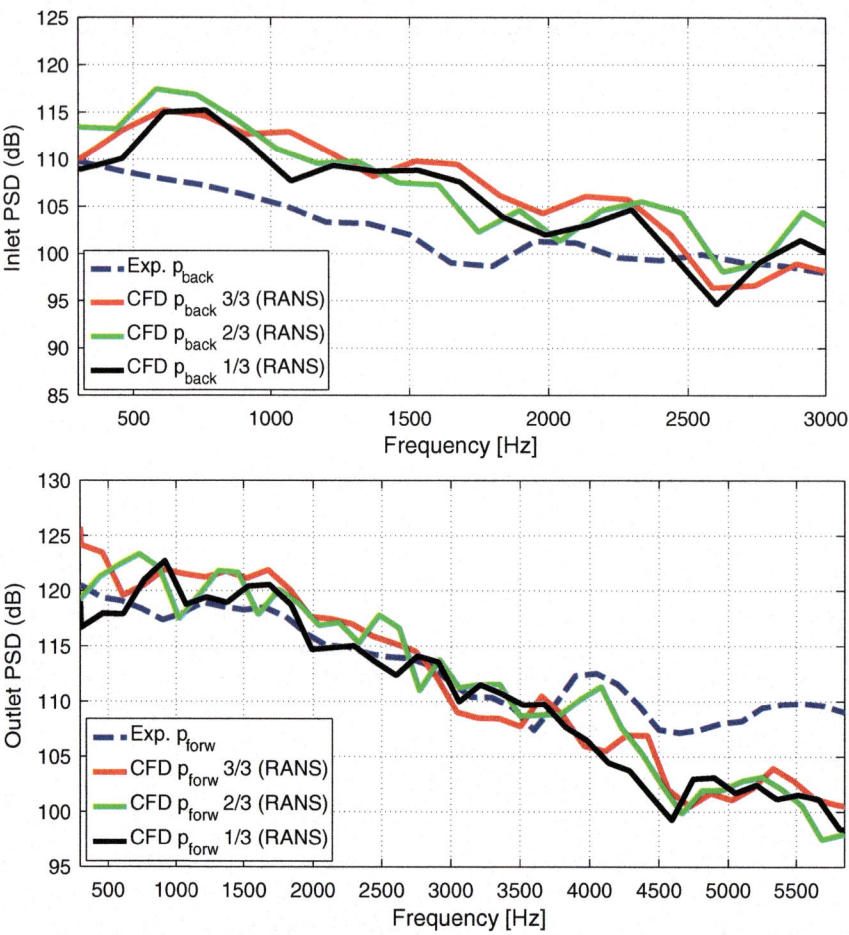

Fig. 3.3 Low frequency PSD of experimental and URANS pressure components at inlet (top) and outlet (bottom) ducts (reprinted from [1], with permission from Elsevier)

Figures 3.3 and 3.4 show that, at the inlet, experimental spectrum at low frequency range decays monotonously. Inlet numerical spectra is predicted by both turbulence approaches alike. Simulations also provide decreasing spectra, but amplitude is over-predicted. Moreover, TCR does not have an impact on PSD.

At the outlet duct, experimental spectrum decreases until 3.5 kHz, presenting a narrow band noise and a constant level afterwards. URANS PSD depicted in Fig. 3.3 fails to predict this change in behavior, showing a constant decay for all TCRs. Conversely, DES spectra (Fig. 3.4) does present a narrow band centered at a frequency from 3.5 to 4 kHz, depending on tip clearance. Then, PSD are constant. The only difference between DES spectrum and experimental one is the broad band that simulations predict from 1.3 to 2.5 kHz. It corresponds to the whoosh noise phenomenon detected at experiments for 77 g/s (see Fig. 2.11).

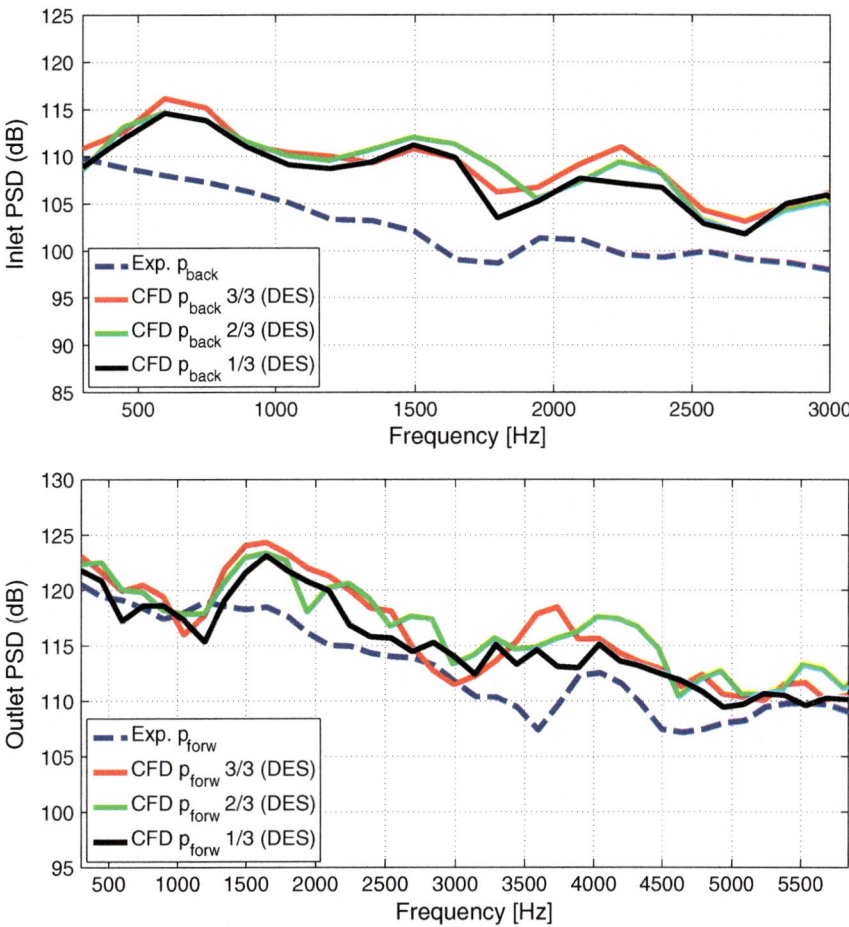

Fig. 3.4 Low frequency PSD of experimental and DES pressure components at inlet (top) and outlet (bottom) ducts (reprinted from [1], with permission from Elsevier)

The differences in spectra prediction between turbulence approaches increase when frequencies above plane-wave range are considered. Figure 3.5 presents spectra provided by URANS simulation against experimental ones. At the inlet, the broadband elevation from 5 to 12 kHz is clearly underpredicted, showing a better agreement in the subsequent decay. PSD at the outlet duct is better predicted, except for frequencies below 10 kHz.

DES (Fig. 3.6) provides an inlet spectrum with a closer estimation of the mean amplitude during the broadband elevation, although it fails to predict PSD decrease above 12 kHz. Narrow band at 4.7–6 kHz is not registered by the experimental measurement. At the outlet duct, PSD is accurately reproduced using the DES. Again, both turbulence models do not predict an influence of TCR on compressor noise at high frequencies (see Figs. 3.5 and 3.6). Conversely, Rolfes et al.

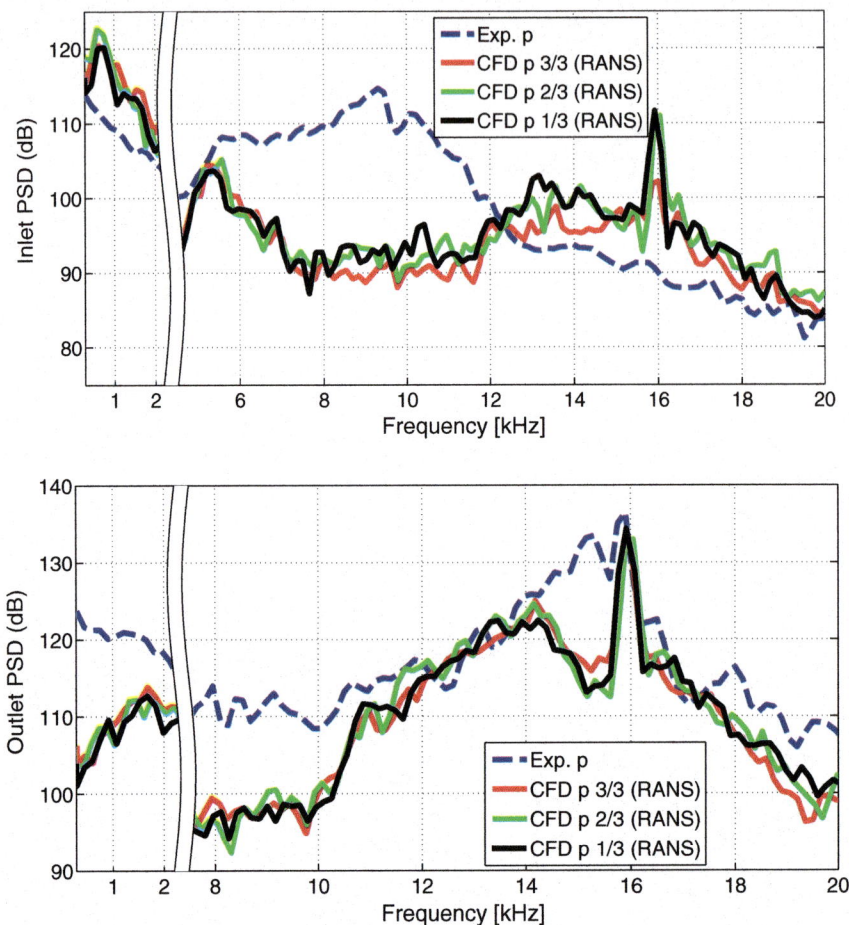

Fig. 3.5 High frequency PSD of experimental and URANS pressure at inlet (top) and outlet (bottom) ducts (reprinted from [1], with permission from Elsevier)

[14] did notice a rise of PSD amplitude in frequencies below BPF with increasing tip clearance size in their experimental measurements of an axial compressor working at near-stall conditions. Also, Ju et al. [15] conducted numerical simulations of an unshrouded centrifugal compressor and found additional dominant frequencies when the tip clearance was reduced with respect to the baseline geometry.

To sum up, Figs. 3.3, 3.4, 3.5 and 3.6 show that DES is able to predict the outlet PSD inflection at 3.5 kHz while URANS is not. Besides, DES presents a greater agreement at high frequencies. Only inlet spectrum above 12 kHz is overpredicted, and it does not represent a major issue since inlet PSD is significantly lower than outlet one (average of 10 dB), and human hearing sensitivity to 12 kHz onwards is limited. Monier et al. [16] showed that RANS models represent properly the

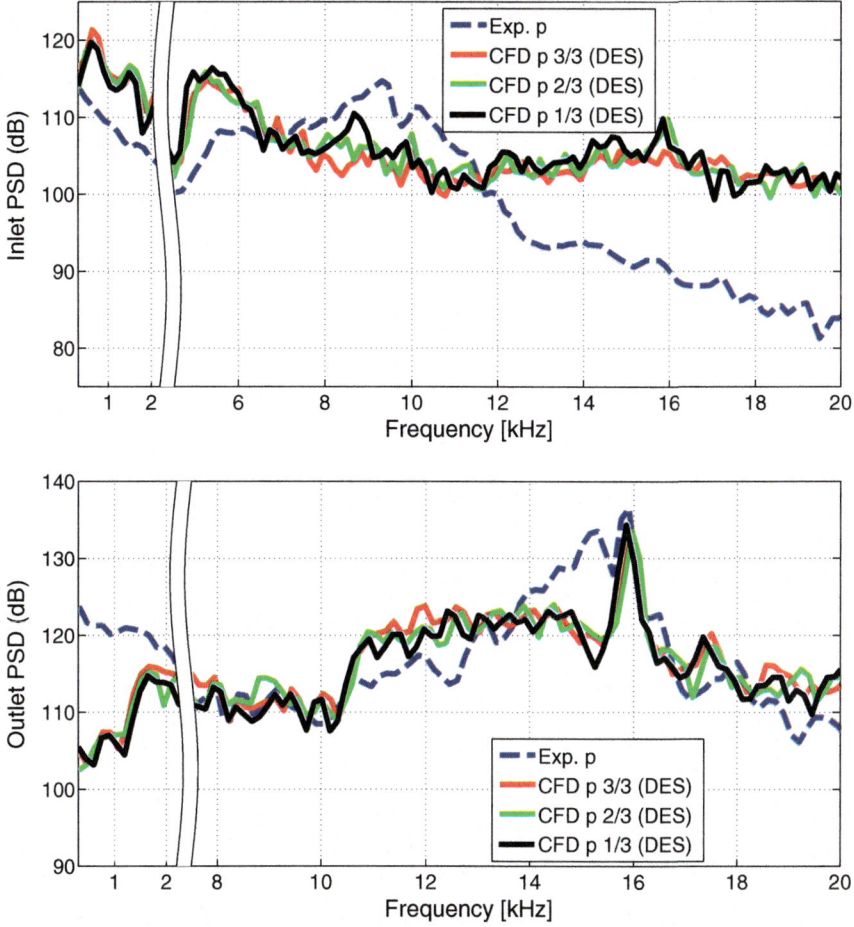

Fig. 3.6 High frequency PSD of experimental and DES pressure at inlet (top) and outlet (bottom) ducts (reprinted from [1], with permission from Elsevier)

time-averaged tip clearance flow but strongly underestimate the Reynolds stresses when compared to a LES approach.

3.6 Flow Field Investigation

In this section, flow field is investigated to determine why TCR is not affecting compressor acoustics in these near-surge operating conditions (mass flow rate is 1.4 times surge mass flow at the speed line of 160 krpm). Only detached-eddy simulations are analyzed, since Sect. 3.5.2 has confirmed their (slightly) superior performance over URANS in terms of noise prediction.

Figure 3.7 shows the mean flow field for the cases with different tip clearance. The different fluid variables have been time-averaged during more than 10 impeller rotations. Line Integral Convolution (LIC) [17] is used to represent relative velocity vectors as an oil flow over isomeridional surfaces. These postprocessing surfaces of revolution are depicted in Fig. 3.2, were created following the approach developed in Sect. A.2.3 and have been flattened by projecting them on a normalized-meridional versus circumferential plane [18]. In Fig. 3.7, LIC is blended with pressure contours.

The impeller rotation is transformed into a horizontal translation from right to left in Fig. 3.7. Therefore, the left side of the blades corresponds to the PS whereas the right side is the SS, as confirmed by pressure contours. The impeller hub is located at

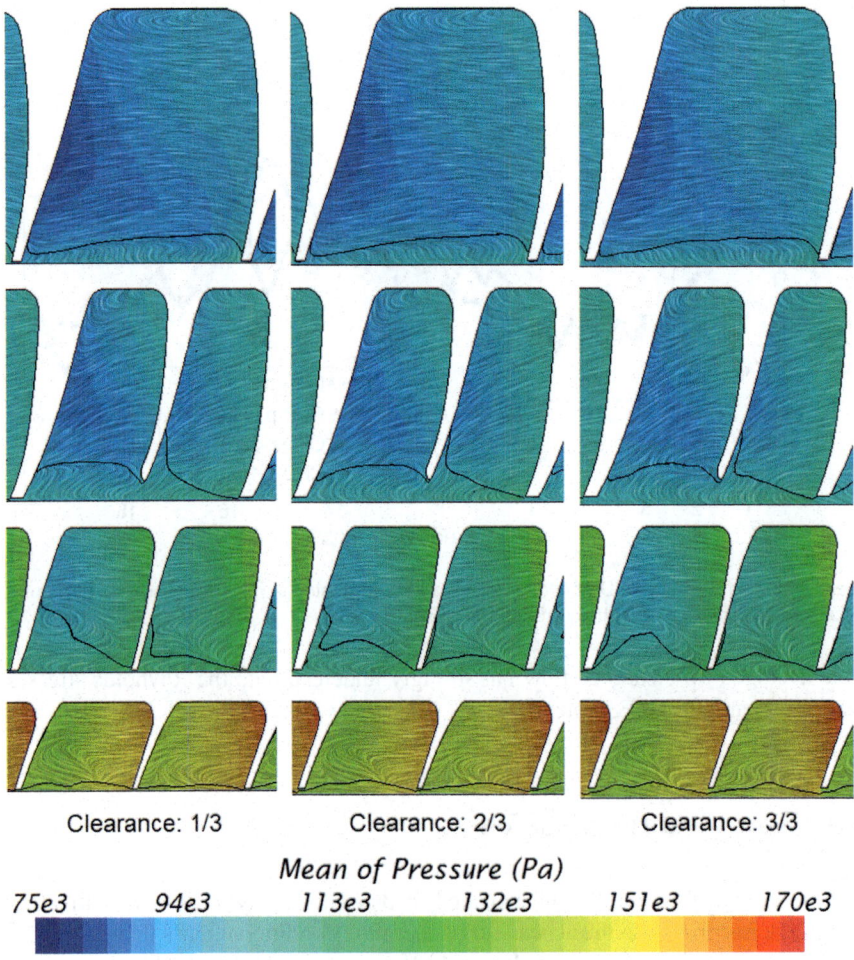

Clearance: 1/3   Clearance: 2/3   Clearance: 3/3

Mean of Pressure (Pa)

75e3   94e3   113e3   132e3   151e3   170e3

Fig. 3.7 LIC of relative velocity vectors colored by pressure at isomeridional surfaces for DES cases with different clearance. Zero meridional velocity is represented as a solid black line (reprinted from [1], with permission from Elsevier)

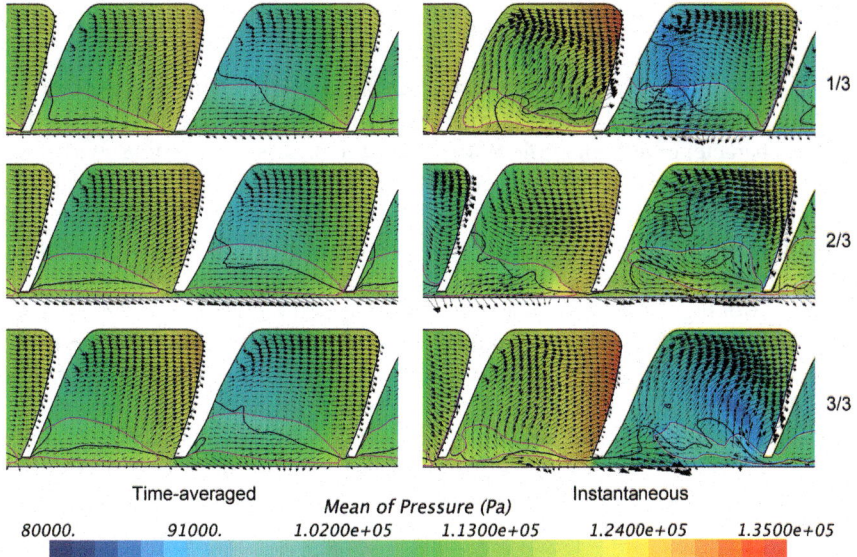

Fig. 3.8 LIC of relative velocity vectors colored by pressure and relative velocity vectors glyph at isomeridional surface number 3 for DES cases with different clearance. Zero meridional velocity is represented as a solid black line whereas zero relative circumferential velocity is represented as a solid violet line (reprinted from [1], with permission from Elsevier)

the upper part of each postprocessing surface whereas the lower part corresponds to the compressor shroud. The solid black line shown in Fig. 3.7 represents a change of sign of meridional component of velocity. The majority of the impeller isomeridional section presents incoming flow, save for the vicinity of the shroud, in which backflow appear. Since tip clearance is included in this recirculating region, tip leakage flow is not only driven by favorable pressure gradient across the blade tip.

In order to shed some light on tip leakage flow, Fig. 3.8 shows the flow field at isomeridional surface number 3 (see Fig. 3.2), for the three cases with different tip clearance. Time-averaged flow field is considered in the left side pictures, whereas the pictures on the right side represent a snapshot of the instantaneous flow. In addition to oil flow colored by pressure and a solid black line indicating zero meridional velocity, as in Fig. 3.7, relative velocity vectors are depicted. It can be seen that some of these vectors indicate flow going to the SS of the passages, i.e., there are cells with relative velocity oriented in the rotational direction. The region in which relative tangential velocity points to the left is highlighted and delimited by a solid violet line in Fig. 3.8.

Left side of Fig. 3.8 shows that the time-averaged flow goes in the streamwise direction and moves from SS to PS at the inner core of the passages (close to the hub). Conversely, flow is going backwards near the shroud as already noted, and a fraction of this recirculating flow goes from passage PS to SS. In the shroud boundary layer, absolute velocity is close to zero, so relative tangential velocity points to the right. These changes in circumferential flow direction results in the creation of secondary flows, such as the vortices observed in Fig. 3.8. Vortex cores are usually regions with local minimum of pressure (see for instance the instantaneous flow for the case with 1/3 of the original clearance).

The air flowing in two opposite directions in the tip clearance that can be observed in the right side of Fig. 3.8 creates a strong shear stress in this region, thus disabling the inviscid character of tip leakage flow [19], at least for near-stall operating conditions. If tip leakage mechanism were mainly inviscid, tip clearance could be resolved with few spanwise nodes (less than 10) [20, 21]. In this work, about 20 cells were used to mesh the spanwise direction of tip clearance (see Fig. 3.2). Van Zante et al. [22] already noted that tip grid can influence prediction of stall inception.

Left side of Fig. 3.8 indicates that mean tip leakage flow moves from the right to the left, and this is justified by the pressure gradient that exists at the different sides of the blade tip. However, if the snapshots at right side of Fig. 3.8 are considered, it can be noticed that instantaneous net tip flow can be oriented in the rotational direction. The reason of the unexpected behavior of tip leakage flow can be explained by Fig. 3.9, which depicts the mean flow in a meridional view at midpitch distance between main blade SS and splitter blade PS (see Sect. A.2.1 for further details on the attainment of

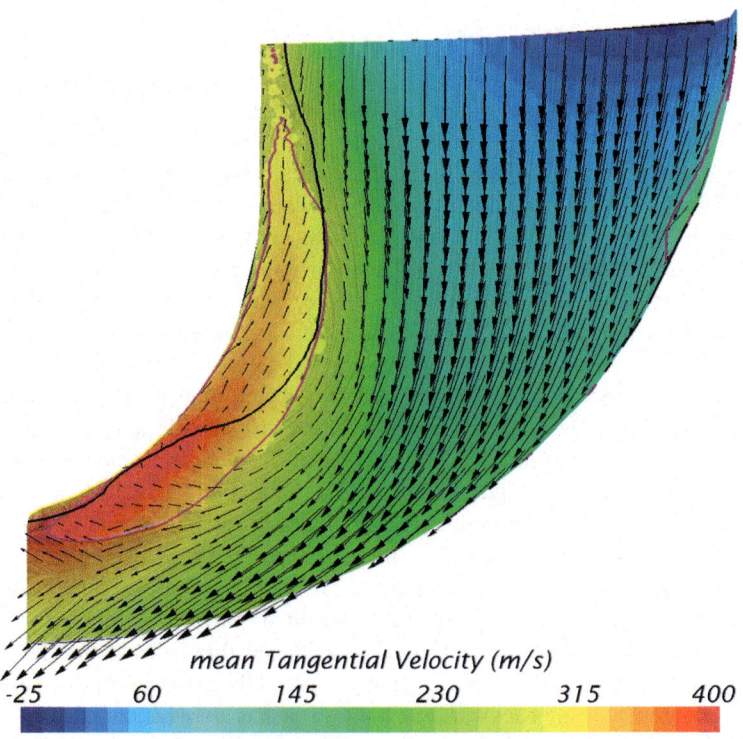

Fig. 3.9 LIC of relative velocity vectors colored by tangential velocity and relative velocity vectors glyph at meridional surface for baseline clearance DES. Zero meridional velocity is represented as a solid black line whereas zero relative circumferential velocity is represented as a solid violet line (reprinted from [1], with permission from Elsevier)

such a postprocessing surface). The solid black line representing change in meridional velocity direction shows that recirculating flow near the shroud almost extends to the trailing edge of the impeller for these operating conditions, although this conclusion could also be drawn from Fig. 3.7. Highest tangential velocities are obtained at the blade trailing edge, since it is the place where blade tip velocity is greatest. The backflow transports the angular momentum that the impeller transmitted to the flow when it was going in the streamwise direction. When the flow is reversed, it will lose some angular momentum because of the transfer to the incoming flow, but also it will also gain tangential velocity, as radius is reduced.

Figure 3.9 shows that mean tangential velocity is in the order of blade tip velocity near the shroud, which explains the existence of regions in which relative velocity is oriented towards the passage SS (marked with a solid violet line in Fig. 3.9). Flow leaking from blade PS to SS through the tip is therefore diminished due to angular momentum of backflows. This fact may explain the reduction of clearance flow rate detected by Liu et al. [23] as operating condition is shifted towards surge.

Some researchers have studied tip clearance noise (TCN), which is attributed to jet-like tip leakage flow, the roll-up of this stream into a vortex and the interaction of these phenomena with adjacent blades [24]. TCN has been thoroughly investigated in axial turbomachinery [25, 26], but it is not so well know in the case of radial turbomachinery. Raitor and Neise [2] reported a narrow band noise in a centrifugal compressor that was considered as TCN because it presented similarities to the case of axial turbomachinery. Bousquet et al. [27] found a periodic vortex shedding due to tip leakage flow in their simulation of a centrifugal compressor at near stall conditions, which may produce a narrow band noise. However, no recirculating flow appeared near the compressor shroud.

In order to produce a narrow ban noise, tip vortex shedding should occur in a freestream flow without any other strong fluctuating phenomena. For the operating conditions studied in this chapter, the whole tip is immersed in the recirculating region, which would mask any noise that may produce tip leakage or even disable the noise source mechanisms. In the simulations it could not be found any periodic tip vortex shedding nor acoustic fluctuations that could be related to tip leakage flow. Only rotating stall was found, but this phenomenon will be studied in detail in Chap. 6. After all, according to Figs. 3.3, 3.4, 3.5 and 3.6, TCR does not affect any acoustic signature. The modification of tip leakage behavior due to the existence of a strong swirling recirculating flow near the shroud is the proposed mechanism to explain that noise generated by the centrifugal compressor is not affected by TCR, at least for the near-stall operating conditions studied in this chapter. Jaatinen-Värr et al. [28] noted that the damage to isentropic efficiency and flow distribution with increasing tip clearance is greater as the working point is shifted towards higher mass flow rates.

3.7 Conclusions

In this chapter, tip clearance sensitivity to flow field and noise generation has been investigated. Measurements of turbocharger shaft motion show that, for a compressor speed of 160 krpm, tip clearance can be modified by a factor of 22–34%. FEM predicts that temperature and rotation cause an additional tip clearance reduction of 11%.

3D CFD simulations with three different tip clearance levels show that clearance reductions cause increases in pressure ratio and isentropic efficiency, but no significant changes in compressor acoustic signature. Two different turbulence approaches (URANS and DES) are studied, and DES is found to provide spectra with better agreement against experimental measurements, particularly at high frequencies (above onset of higher order acoustic modes).

Insensitivity of spectra to TCR is explained by flow field configuration. Since operating condition is close to surge, the flow in the vicinity of the shroud is reversed and tip leakage flow is not only driven by pressure gradient at blade tip. A complicated flow pattern is detected in this region, but the recirculating flow comprises tip clearance. A coherent noise source mechanism due to tip leakage flow cannot be established, so variations of TCR do not have an impact on noise generation. Conversely, compressor performance indicators *are* affected by TCR, because tighter tip clearances lead to smaller backflow regions and reduced mixing losses.

These conclusions are valid for the studied operating conditions. Compressors working with higher mass flow rates should present less backflows close to the shroud and tip sensitivity to noise generation may be different. In any case, the present chapter shows that, for operating conditions close to surge, researchers can rely on CAD tip clearance profile when analyzing acoustic behavior of centrifugal compressors. Computational effort does not allow one to perform tip clearance sensitivity analysis at other operating conditions with higher mass flow rate. Therefore, reverse engineering geometry will be used in the remainder of the book.

Once tip clearance is determined, other model parameters should be investigated so as to select the optimal configuration to predict compressor noise generation. Chapter 4 is devoted to this particular matter.

References

1. J. Galindo, A. Tiseira, R. Navarro, M. López, Influence of tip clearance on flow behavior and noise generation of centrifugal compressors in near-surge conditions. Int. J. Heat Fluid Flow **52**, 129–139 (2015). https://doi.org/10.1016/j.ijheatfluidflow.2014.12.004
2. T. Raitor, W. Neise, Sound generation in centrifugal compressors. J. Sound Vib **314**, 738–756 (2008). https://doi.org/10.1016/j.jsv.2008.01.034. ISSN: 0022-460X
3. F. Mendonça, O. Baris, G. Capon, Simulation of Radial Compressor Aeroacoustics using CFD, in *Proceedings of ASME Turbo Expo 2012. GT2012-70028* (ASME, 2012), pp. 1823–1832. https://doi.org/10.1115/GT2012-70028

4. I. Tomita, S. Ibaraki, M. Furukawa, K. Yamada, The effect of tip leakage vortex for operating range enhancement of centrifugal compressor. J. Turbomach. **135**(5), 8 (2013). https://doi.org/10.1115/1.4007894

5. S.N. Danish, M. Chaochen, Y. Ce, L. Wei, Comparison of two methods to increase the tip clearance and its effect on performance of a turbocharger centrifugal compressor stage, in *International Conference on Energy and Environment* (2006), p. 7

6. Y. Jung, M. Choi, S. Oh, J. Baek, Effects of a nonuniform tip clearance profile on the performance and flow field in a centrifugal compressor. Int. J. Rotating Mach. **2012** (2012). https://doi.org/10.1155/2012/340439

7. H.-L. Wang, G. Xi, J.-Y. Li, M.-J. Yuan, Effect of the tip clearance variation on the performance of a centrifugal compressor with considering impeller deformation. Proc. Inst. Mech. Eng. Part A J. Power Energy **225**(8), 1143–1155 (2011). https://doi.org/10.1177/0957650911416914

8. F. Nelson, Rotordyamics without equations. Int. J. COMADEM **10**(3), 2–10 (2007)

9. T. Sotome, S. Sakoda, Development of manufacturing technology for precision compressor wheel castings for turbochargers. Furukawa Rev. **32**, 56–59 (2007), https://www.furukawa.co.jp/review/fr032.htm

10. J.V. Pastor, J.R. Serrano, V. Dolz, M.A. López, F. Bouffaud, Study of turbocharger shaft motion by means of non-invasive optical techniques: application to the behaviour analysis in turbocharger lubrication failures. Mech. Syst. Sig. Proc. **32**, 292–305 (2012). https://doi.org/10.1016/j.ymssp.2012.04.020

11. J.V. Pastor, J.R. Serrano, V. Dolz, M.A. López, A non-invasive optical technique to observe turbocharger shaft whirl, in *XXI Biennial Symposium on Measuring Techniques in Turbomachinery* (2012). ISBN:978-84-8363-966-5

12. J. Galindo, J.R. Serrano, V. Dolz, M.A. López, Behavior of an IC engine turbocharger in critical condition of lubrication. SAE Int. J. Engines **6** (2013). https://doi.org/10.4271/2013-01-0921

13. M. López, *Estudio Teórico-Experimental de la Dinámica Rotacional de Turbocompresores de MCIA. Aplicación al diagnóstico de fallos*. Ph.D. thesis. Universitat Politècnica de València (2014)

14. M. Rolfes, M. Lange, R. Mailach, Investigation of performance and rotor tip flow field in a low speed research compressor with circumferential groove casing treatment at varying tip clearance. Int. J. Rotating Mach. **2017**, 14 (2017). https://doi.org/10.1155/2017/4631751

15. W. Ju, S. Xu, X. Wang, X. Chen, S. Yang, J. Meng, Numerical study on the influence of tip clearance on rotating stall in an unshrouded centrifugal compressor, in *ASME Turbo Expo 2017: Turbomachinery Technical Conference and Exposition. GT2017-64452* (2017), p. 8. https://doi.org/10.1115/GT2017-64452

16. J.-F. Monier, J. Boudet, J. Caro, L. Shao, Turbulent energy budget in a tip leakage flow: a comparison between RANS and LES, in *ASME Turbo Expo 2017: Turbomachinery Technical Conference and Exposition* (American Society of Mechanical Engineers, 2017), p. 10

17. B. Cabral, L.C. Leedom, Imaging vector fields using line integral convolution, in *Proceedings of the 20th Annual Conference on Computer Graphics and Interactive Techniques* (ACM, 1993), pp. 263–270. https://doi.org/10.1145/166117.166151

18. M. Drela, H. Youngren, A User's Guide to MISES 2.63. MIT Aerospace Computational Design Laboratory (2008)

19. J.A. Storer, N.A. Cumpsty, Tip leakage flow in axial compressors. J. Turbomach. **113**(2), 252–259 (1991). https://doi.org/10.1115/1.2929095

20. H.-J. Eum, Y.-S. Kang, S.-H. Kang, Tip clearance effect on through-flow and performance of a centrifugal compressor. English. KSME Int. J. **18**(6), 979–989 (2004). https://doi.org/10.1007/BF02990870. ISSN:1738-494X

21. Y.K.P. Shum, N.A. Cumpsty, C.S. Tan, Impeller-diffuser interaction in a centrifugal compressor. J. Turbomach. **122**(4), 777–786 (2000). https://doi.org/10.1115/1.1308570

22. D.E. Van Zante, M.D. Hathaway, T.H. Okiishi, A.J. Strazisar, J.R. Wood, Recommendations for achieving accurate numerical simulation of tip clearance flows in transonic compressor rotors. J. Turbomach. **122**(4), 733–742 (2000). https://doi.org/10.1115/1.1314609

23. Z. Liu, Y. Ping, M. Zangeneh, On the nature of tip clearance flow in subsonic centrifugal impellers. English. Sci. China Technol. Sci. **56**(9), 2170–2177 (2013). https://doi.org/10.1007/s11431-013-5313-3. ISSN:1674-7321

24. M.C. Jacob, J. Grilliat, R Camussi, G.C. Gennaro, Aeroacoustic investigation of a single airfoil tip leakage flow. Int. J. Aeroacoustics **9**(3), 253–272 (2010)

25. T. Fukano, C.-M. Jang, Tip clearance noise of axial flow fans operating at design and off-design condition. J. Sound Vib. **275**(3), 1027–1050 (2004). https://doi.org/10.1016/S0022-460X(03)00815-0

26. F. Kameier, W. Neise, Experimental study of tip clearance losses and noise in axial turbomachines and their reduction. J. Turbomach. **119**(3), 460–471 (1997)

27. Y. Bousquet, X. Carbonneau, G. Dufour, N. Binder, I. Trebinjac, Analysis of the unsteady flow field in a centrifugal compressor from peak efficiency to near stall with full-annulus simulations. Int. J. Rotating Mach. **2014**, 11 (2014). https://doi.org/10.1155/2014/729629

28. A. Jaatinen-Värri, T. Turunen-Saaresti, A. Grönman, P. Röyttä, J. Backman, The tip clearance effects on the centrifugal compressor vaneless diffuser flow fields at off-design conditions, in *10th European Turbomachinery Conference* (2013)

Chapter 4
Sensitivity of Compressor Noise Prediction to Numerical Setup

4.1 Introduction

So far, a CFD model of a centrifugal compressor has been implemented in Star-CCM+ [1], taking as a reference the setup used by Mendonça et al. [2]. A methodology to validate the model against experimental measurements was developed in Sect. 2.4, so the baseline configuration can be interrogated. A first result is that the geometry obtained by reverse engineering can be used *as is*, because tip clearance size does not have a significant impact on compressor noise.

In this chapter, main decisions on numerical setup are either justified or evaluated in terms of noise generation. First, wheel rotation strategies are discussed in Sect. 4.2. Then, Sect. 4.3 is devoted to the description of the different turbulence models that could have been used, eventually selecting URANS and DES approaches, that will be employed in the remainder of the book. After that, the baseline setup is analyzed in terms of compressor performance variables and noise production. Grid spacing (Sect. 4.4), solver type (Sect. 4.5) and times-step size (Sect. 4.6) are evaluated. As in the case of tip clearance (see Chap. 3), sensitivity analyses performed in this chapter are carried out for 60 g/s working point. These near-surge conditions are expected to be the most sensitive to numerical configuration.

4.2 Wheel Rotation Approach

One of the capital choices when defining the set-up in a turbomachinery CFD case is the approach to simulate impeller motion. Two different methods stand out: frozen rotor and rigid body motion (RBM). In any case, interfaces are used to split the rotating region (rotor) from the stationary zones (volute and ducts).

© Springer International Publishing AG 2018
R. Navarro García, *Predicting Flow-Induced Acoustics at Near-Stall Conditions in an Automotive Turbocharger Compressor*, Springer Theses, https://doi.org/10.1007/978-3-319-72248-1_4

Frozen rotor approach, also known as multiple reference frame (MRF), is based on the definition of a coordinate system that rotates rigidly with the impeller. Inside the rotor region, the equations are solved in the rotational reference frame and Coriolis and centrifugal forces are introduced in the Navier/Stokes equations as source terms. Conversely, stationary zones use a non-rotational coordinate system. With this method, the impeller does not actually rotate, although Zheng et al. [3] proved that the particular position of the impeller regarding the volute tongue presents only a slight impact on the numerical predictions.

MRF approach allows the use of steady simulations just by neglecting the transient terms in the original unsteady equations, thus reducing the computational effort by a order of magnitude when compared to transient simulations [4]. Hillewaert and Van den Braembussche [5] stated the frozen rotor model was not the most appropriate for radial compressor simulations. Similar behavior was found in the work of Liu and Hill [6], in which the error induced by the use of the MRF approach was evaluated for different compressor geometries, analyzing in this way the effect of the rotor-stator interaction. Zheng et al. [3] employed the frozen rotor approach to investigate the flow inside the volute of a centrifugal compressor working at off-design conditions. Comparison with test bench results in terms of flow capacity showed excellent agreement for low rotational speed, with decreasing accuracy when increasing the rotational speed. At highest considered rotational speed, peak pressure ratio was overestimated more than 10%. Zheng et al. [3] indicated that the frozen rotor approach has only limited capability to predict the flow field in the impeller. They acknowledged the necessity of carrying unsteady simulations for further design steps.

Besides, turbomachinery can also be modeled by performing unsteady simulations in which the impeller region is actually rotated at each time step. Connectivity of cells on either side of the interface should be therefore changed at every time step, so sliding interfaces are used in a RBM simulation. The increase of computational capabilities in the last years has increased the popularity of RBM simulations, which shows good agreement against experimental measurements in both radial turbines [7, 8] and centrifugal compressors [9, 10]. Currently, steady state calculations with frozen rotor method are only used when several cases have to be performed, due to the reduced computational cost of each simulation. In this way, Jiao et al. [11] used the MRF approach and a mixing plane interface (in which flow field data is spatially averaged) to analyze the performance of a dual volute compressor.

The ultimate goal of this book is the analysis of compressor aeroacoustics, so unsteady flow effects cannot be neglected and transient simulations should be performed. Unsteady calculations could be carried out using frozen rotor approach instead of RBM, but the save in computational effort would be small and transient effects such as blade passing tonal noise would not be predicted. In this work, rigid body motion approach is therefore used.

4.3   Turbulence Models

Ideally, one would like to calculate the centrifugal compressor flow field by means of a Direct Numerical Simulation (DNS), in which "the unsteady Navier/Stokes equations are solved on spatial grids that are sufficiently fine that they can resolve the Kolmogorov length scales at which energy dissipation takes place and with time steps sufficiently small to resolve the period of the fastest fluctuations" [12]. Currently, this is only feasible for simple cases (pipes, channels, etc.) that are run using the state-of-the-art high-performance computers due to the outrageous mesh requirements that a DNS dictates [13]. These simulations are useful to improve the understanding of turbulence physics (see for instance the work of Hoyas and Jiménez [14]). DNS approach cannot be employed for *industrial* cases, so that turbulence needs to be modeled, which constitutes a major issue in CFD simulations [15]. In order to predict compressor aeroacoustics, three approaches to model turbulence could be considered: Unsteady Reynolds Averaged Navier Stokes Simulations (URANS), Detached Eddy Simulation (DES) or Large Eddy Simulation (LES).

In LES [16, 17] the large turbulent eddies are resolved, modeling only the small scales by means of a sub-grid scale (SGS) model. This approach takes advantage of the fact that small scales are usually quite isotropic, because macroscopic flow behavior has a limited effect on small scales. LES are often used in computational aeroacoustics (CAA) [18], due to their potential to capture broadband noise turbulent sources. Dufour et al. [19] reviewed some works using large eddy simulations for predicting turbomachinery flow. By looking at their survey, it can be concluded that LES is much more extended in axial turbomachinery than in their radial counterparts. To the authors's knowledge, LES in turbocharger compressors have only been covered by Karim et al. [20] and the set of works performed by researchers from KTH and University of Cincinnati. Hellström et al. [21, 22], Jyothishkumar et al. [23] and Semlitsch et al. [9] reported good agreement between their large eddy simulations and experimental measurements, showing the potential of LES to predict centrifugal compressor flow behavior, particularly at near-surge conditions.

RANS approach is obtained by performing the so-called Reynolds Averaging over Navier/Stokes equations. In this way, an instantaneous variable is decomposed into a time-averaged value plus a fluctuating quantity. Averaged flow variables are governed by RANS equations whereas fluctuating terms should be modeled. In URANS, mean values are unsteady, which may seem a paradox. Ensemble averaging would deliver a mean quantity as "(...) the average of the instantaneous values of the property over a large number of repeated identical experiments" [12]. This process can also be considered as a time averaging over a turbulent time scale, which is assumed to be smaller than mean flow time scale. URANS will therefore perform poorly in cases presenting turbulent and mean flow time scales of the same order.

In any case, the key feature in a RANS method is the modeling approach for the Reynolds stresses defined by the fluctuating terms. These new unknowns require a closure model (usually referred directly as RANS turbulence model) to close the system of equations. Researchers have developed and tested a great number of different RANS turbulence models. The works of Aghaei et al. [24], Borm et al. [25] and Mangani et al. [26] are an excerpt of papers comparing different closure models in centrifugal compressors.

In this book, URANS simulations are performed with $k - \omega$ SST model [27]. This model is very suited for compressors, for its accuracy in predicting flow with adverse pressure gradients. Menter et al. [28] applied the SST model for different turbomachinery cases, for which a good agreement against experimental measurements was found. The model proved to be good at capturing usual turbomachinery flow features such as flow separation, swirl or strong flow mixing. Robinson et al. [29] also used the aforementioned turbulence model to analyze the influence of the spacing between impeller and vaned diffuser in a centrifugal compressor. Smirnov et al. [30] applied the $k - \omega$ SST model for the calculation of flow in a one-stage centrifugal compressor with a vaned diffuser. Good agreement of the computed velocity field with the measurements data at the impeller exit was obtained. Downstream of the diffuser vane, prediction quality depended on the operating point. A comparison between turbulence models was done by Smirnov et al. [30] at one operating point. The SST model provided better agreement with experimental data than $k - \epsilon$ and $k - \omega$ models. Pinto et al. [31] performed a review of the state of the art in turbomachinery CFD simulations from which one can realize that $k - \omega$ SST model is employed more frequently than $k - \epsilon$ for centrifugal compressors. In fact, contributions using $k - \epsilon$ are frequently the oldest references in the review conducted by Pinto et al. [31].

Finally, DES model is an hybrid RANS-LES approach in which the boundary layer is modeled using URANS and LES is employed for the free stream detached flow. With such a blending between turbulence approaches, DES is intended to save computational cost when compared with LES but retaining some of the eddy-resolving capabilities, with the corresponding improvement in broadband noise predictions regarding URANS. However, it is still not very popular in turbomachinery simulations. Tucker [32] listed 40 turbomachinery cases with eddy-resolving turbulence models, only 3 of them being hybrid RANS-LES.

DES are conducted in this work in combination with a SST $k - \omega$ turbulence model, "which functions as a sub-grid-scale model in regions where the grid density is fine enough for a large-eddy simulation, and as a Reynolds-averaged model in regions where it is not" [33]. Mendonça et al. [2] showed that low-Reynolds number near-wall resolution improved the prediction of the compressor map compared to experimental measurements when using steady-state RANS $k - \omega$ SST simulations. In order to predict broadband aeroacoustic excitations, they therefore decided to use a DES approach, with a low-Reynolds number $k - \omega$ SST turbulence model.

Particularly, IDDES [34, 35] is used, which combines WMLES and DDES hybrid RANS-LES approaches. Mockett [36] studied several DES models, concluding that IDDES extended the range of suitable applications of DDES, by reducing the issue with log-layer mismatch (LLM). The meshes used in this book may not be fine enough to take advantage of the WMLES branch in IDDES, but at least this turbulence model should perform as DDES. Van Rennings et al. [37] used DDES for simulations of an axial compressor near stall. Imposed geometrical periodicity was found to affect large scale turbulent structures, concluding that the full compressor should be modeled for the sake of accuracy.

Since choice of URANS or DES turbulence approach may have an impact on the predicted compressor mean flow and aeroacoustics, both models have been used throughout this book. In Chap. 3, these two models were used for the tip clearance sensitivity analysis at 60 g/s, showing superior performance of DES over URANS in PSD at certain frequencies. In this chapter, the effect of mesh density, time-step size and solver type is going to be assessed with both URANS and DES calculations. Finally, Chap. 6 will display PSD at three operating conditions, using again URANS as well as DES.

4.4 Mesh

In order to model the full 3D geometry of a centrifugal compressor, one has to determine the extension of the domain (length of inlet and outlet ducts), select the type of cells of the mesh, the amount of elements that can be used so as to keep an adequate workaround time taking into account the available computational resources, whether to resolve the tip clearance or not, etc.

One of the many aspects of the mesh that should be also evaluated is the wall resolution. Turbulence models present different wall treatment depending on the values of the dimensionless wall distance, y^+, provided by the cell contiguous to the wall.

y^+ is calculated as follows:

$$y^+ = \frac{y u_\tau}{\nu},\tag{4.1}$$

where y is the distance of the cell's centroid to the wall, u_τ is the friction velocity and ν is the kinematic viscosity. y^+ is an important quantity in turbulence because it determines the limits of the multi-layer structure of the dimensionless velocity u^+ known as "law of the wall". For more details, the reader can refer to the book of Veersteg and Malalasekera [12].

When $y^+ < 5$, the cell is said to be in the viscous sublayer, where the flow is laminar. If first y^+ is close to one and there are several cells in this region, the code attempts to resolve the flow in this zone, provided that the turbulence model allows this feature. When $y^+ > 30$, wall functions are used, in which velocity is determined by the log-law. If y^+ lays in the intermediate region, known as the buffer layer, there is not a direct approach to obtain the velocity, so it is usually obtained by blending extrapolations of the other regions. With SST $k - \omega$ turbulence model, the flow can be resolved up to the viscous sublayer. Moreover, StarCCM+ [1] allows the use of the so-called "all y^+ wall treatment", that will perform as a wall function or resolve the viscous sublayer, depending on the wall y^+. Besides, it claims to produce reasonable results for cells laying in the buffer region.

As commented in Sect. 4.1, the work of Mendonça et al. [2] has been used to define
the baseline set-up for Chaps. 2 and 3. Particularly, a mesh of 9.5 million polyhedral
cells has been used so far. The compressor model includes blade tip clearance and
backplate region, together with 5-diameters-long inlet and outlet ducts. In this section,
two additional meshes are studied: one with half of the elements of the baseline case
(4 million) and another one twice as fine (19 million cells).

A literature survey has been performed, providing a brief description of the grids
used in several recent CFD simulations of centrifugal compressors:

- Mendonça et al. [2] employed a 9 million cells (probably polyhedral) mesh, which
  presented $y^+ \approx 1$ on the rotor blades. The grid is fine enough to resolve the
  tip-clearance. The domain included inlet and outlet ducts of 0.5 m.
- Guo et al. [38] and Chen et al. [39] adopted the same grid. Structured grid was
  used for rotor and diffuser flow passages. The O-type mesh was employed near
  the rotor blades, and mean y^+ value was about 20. The H-type grid was applied
  to the other parts of the passages. An independent H-type mesh was employed in
  the rotor tip clearance. The total number of the grid cells for the passages is over
  1.6 millions. For the volute, an unstructured tetrahedral mesh was generated with
  refinement around the tongue region. A structured grid was employed at the volute
  exit. The total number of the volute mesh elements is over 0.8 millions.
- Hellström et al. [21] employed a grid with approximately 11.7 million cells, with
  predominantly hexahedral elements. 4.1 million cells are located in the wheel
  region and 7.6 million in the inlet region and the compressor housing. To capture
  the boundary layer, the mesh is refined at the walls.
- Lee et al. [40] used a polyhedral grid with inlet and outlet ducts of about 0.4 m.
  The number of elements employed by the grid is not specified.
- Jiao et al. [11] applied a grid containing 2 million elements to simulate a dual
  volute compressor working as a single volute one. The impeller and the diffuser
  are meshed with hexahedral elements, while volute is meshed with tetrahedral
  ones. The clearance gap between the blades and the shroud is resolved using 4
  cells. Near the solid walls, the range of y^+ is between 31 and 290.
- Robinson et al. [29] adopted a grid with 6.5 million elements when performing
  transient computations of the full 3D geometry of a centrifugal compressor. A
  mesh independence study showed that the selected grid provided values of efficiency with a difference lower than 0.2% compared with a mesh twice as fine. The
  clearance gap seems to be resolved, although no details of the amount of cells used
  are provided. An average y^+ value of 4.5 is employed near the walls.
- Zheng et al. [3] applied a structured grid with 3.5 million cells to predict performance of a centrifugal compressor. The boundary layer resolution keeps y^+ values
  between 1 and 5. The tip clearance is meshed. A butterfly-type mesh is applied to
  the volute.
- Smirnov et al. [30] used a hexahedral grid with 2.7 million elements for the computation of a full 360° model of a centrifugal compressor with vaned diffuser.

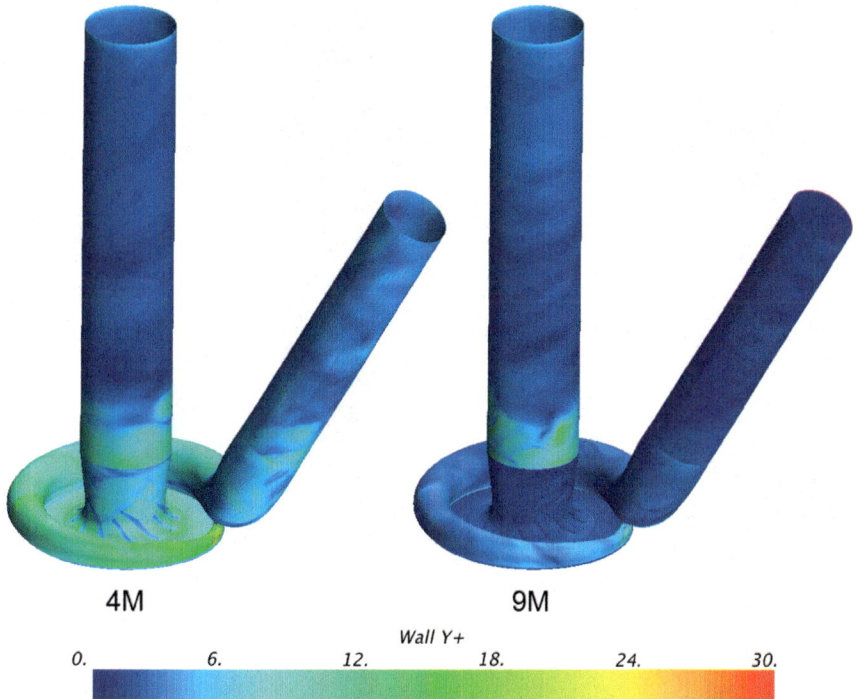

Fig. 4.1 Global wall y^+ contours at 4 million and 9 million meshes simulated with DES

- Tomita et al. [41] employed a grid containing 3.1 million cells to compute two centrifugal compressors. 2.5 million hexahedral elements were used at the impeller, in which tip clearance was considered.

The literature review shows that the set of three meshes studied in this section present an average to above-average rating in terms of element quantity (4–19 million cells) for centrifugal compressor simulations. First, wall y^+ will be investigated. Since wall y^+ does not vary significantly between URANS and DES equivalent cases, y^+ figures will only correspond to the latter.

Figure 4.1 depicts y^+ values at global domain walls. Only the two coarser meshes are considered. The scale is set with a maximum of $y^+ = 30$ and no walls are in warm colors, so wall functions are not used in the simulations. The case with 4 million cells presents all walls except inlet and outlet ducts in the buffer region. Mesh refinement performed to obtain 9M grid provides a reduction in wall y^+ values, although the inlet duct next to the shroud lays also in the buffer zone.

Figure 4.2 displays wall y^+ contours with a tighter scale (max. y^+ is 3) for cases with 9 and 19 million elements. The finest mesh presents y^+ values less than 1.5, thus providing a good resolution of the viscous sublayer. Case with 9M cells should be able to resolve the viscous sublayer in the shroud, diffuser and outlet duct. Most of the volute and the last part of the inlet duct will lay in the buffer region.

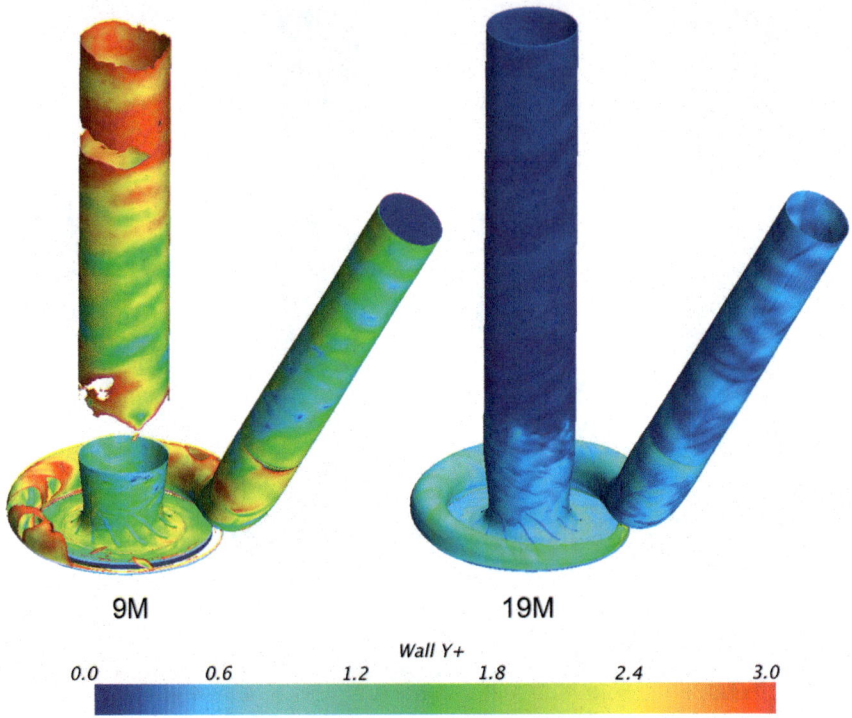

Fig. 4.2 Global wall y^+ contours at 9 million and 19 million meshes simulated with DES

Wall y^+ distribution in the impeller is represented in Fig. 4.3. Impellers on the left hand side are depicted with an upper scale limit of $y^+ = 30$ whereas this limit is reduced up to $y^+ = 3$ in the right hand side. Again, no wall functions are used in the impeller blades, not even for the coarsest mesh. Case with 4 million cells cannot resolve the viscous sublayer, thus laying in the buffer zone. Conversely, the two finer meshes should be able to resolve the viscous sublayer, although case with 19 million elements may provide superior performance.

Taking into account the differences in wall resolution (among other aspects) observed for the studied meshes, their predictions in terms of global variables and noise is evaluated. Table 4.1 includes compressor performance variables calculated using Eqs. 2.3 and 2.4 for the different grids. Relative difference against experimental measurements (Eq. 2.5) and relative difference against baseline case (9 million cells) with the corresponding turbulence model, defined by

$$\epsilon_{base}(\%) = \frac{\phi_{CFD} - \phi_{CFD,base}}{\phi_{CFD,base}} \cdot 100, \tag{4.2}$$

are also included in Table 4.1.

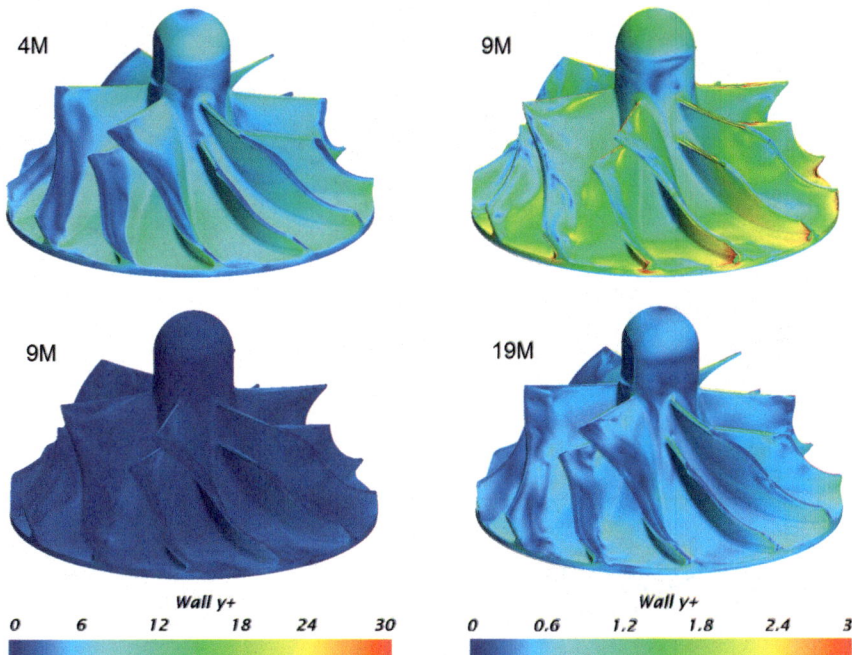

Fig. 4.3 Impeller wall y^+ contours at cases with different meshes simulated with DES

Table 4.1 Experimental and numerical compressor global variables for mesh sensitivity analysis at 60 g/s

Case		$\Pi_{t,t}$ [–] ($\phi/\epsilon_{exp}/\epsilon_{base}$)	W_u [kJ · kg^{-1}] ($\phi/\epsilon_{exp}/\epsilon_{base}$)	η_s [%] ($\phi/\epsilon_{exp}/\epsilon_{base}$)
Exp.		2.22/–/–	121/–/–	62.2/–/–
URANS	4M	2.18/−2.1/−1.5	126/3.7/3.0	59.2/−4.9/−5.0
	9M	2.21/−0.5/–	122/0.7/–	62.3/0.1/–
	19M	2.21/−0.7/−0.1	122/0.4/−0.3	62.4/0.2/0.1
DES	4M	2.17/−2.2/−2.6	125/3.2/2.6	59.4/−4.6/−6.1
	9M	2.23/0.4/–	122/0.6/–	63.2/1.6/–
	19M	2.23/0.5/0.1	122/0.1/−0.5	63.6/2.2/0.6

In terms of global variables, doubling the mesh size from 9 to 19 million elements is not worth the increase in computational effort, because a relative error below 1% is obtained for both turbulent models. Conversely, cases with 4 million cells underpredict isentropic efficiency by more than 5% regarding 9 million elements case. The proximity of the operating point to surge may have increased the sensitivity of the global variables to the mesh, because the literature survey show that 4 million cell meshes are commonly used.

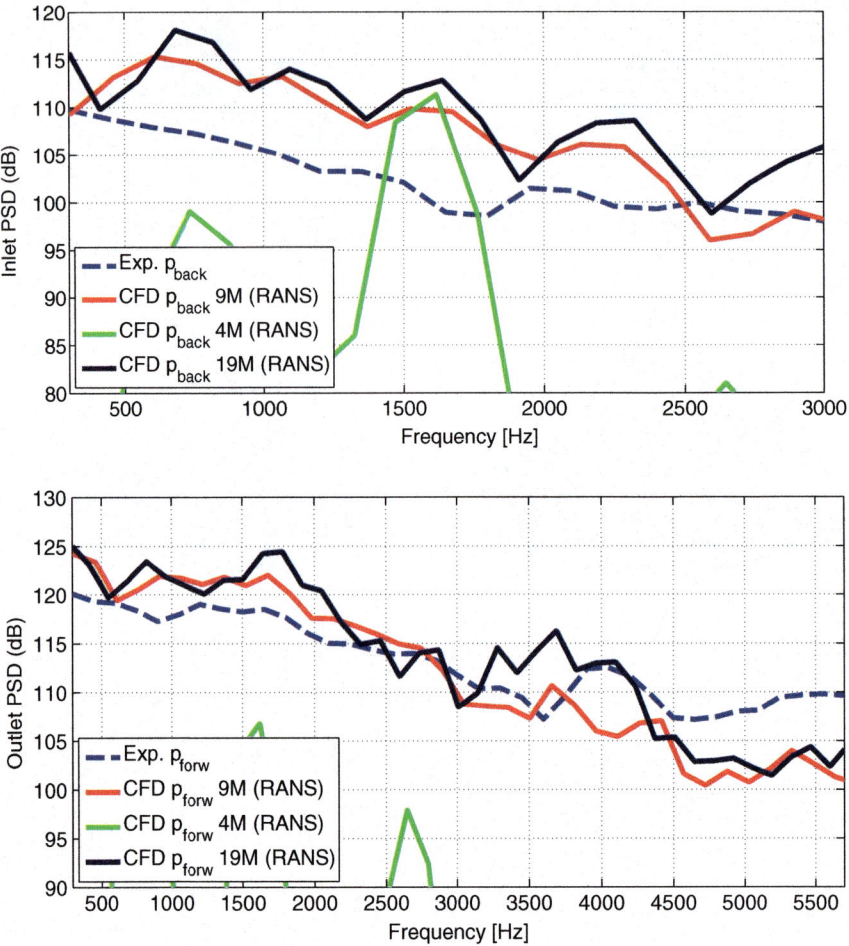

Fig. 4.4 Low frequency PSD of experimental and numerical pressure components at inlet (top) and outlet (bottom) ducts at 60 g/s for URANS simulations with different meshes

Even though Table 4.1 proves that 9 million cells are enough to predict compressor performance variables with little error, mesh size should be assessed in terms of noise prediction. Figures 4.4, 4.5, 4.6 and 4.7 show PSD for different meshes using URANS and DES.

Figures 4.4, 4.5, 4.6 and 4.7 prove that PSD predicted by 9 million elements grid do not significantly differ from the ones obtained using 19 million cells. Only inlet duct spectra at high frequency (top of Figs. 4.5 and 4.7) present some differences, especially in URANS simulations, but the slight change in PSD is not worth the increase in computational time. However, the most significant conclusion obtained by observing Figs. 4.4, 4.5, 4.6 and 4.7 is that URANS simulation of 4 million grid fails

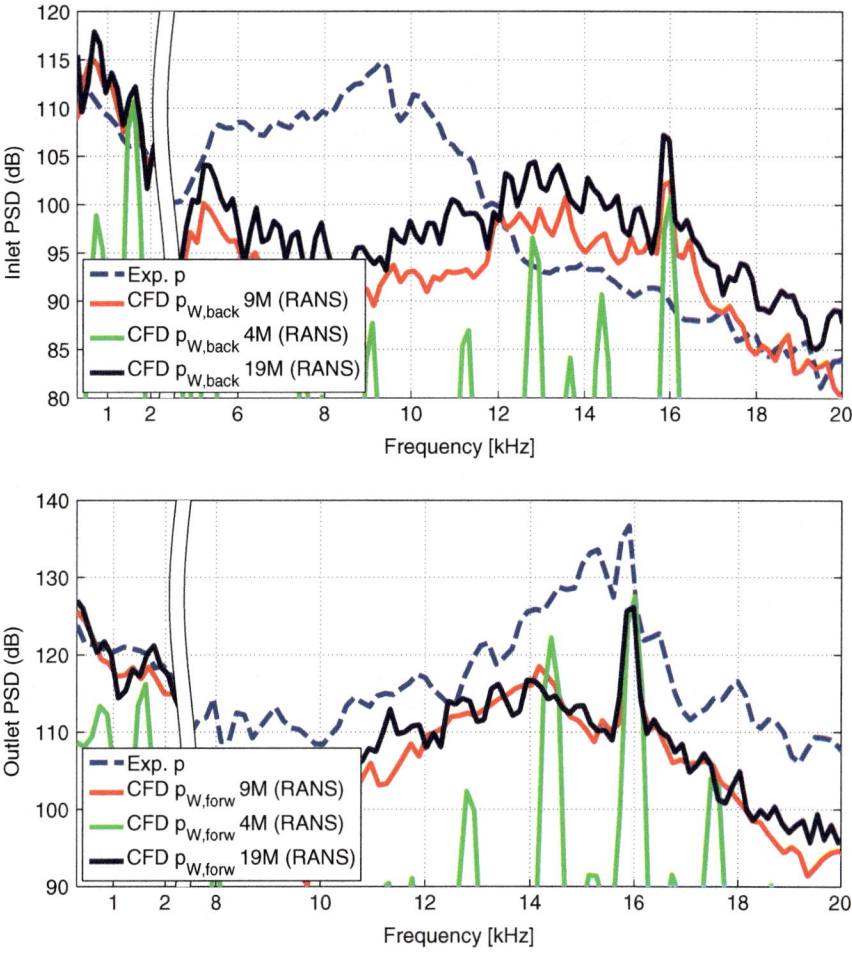

Fig. 4.5 High frequency PSD of experimental and numerical pressure components at inlet (top) and outlet (bottom) ducts at 60 g/s for URANS simulations with different meshes

to predict compressor noise generation, and particularly mean broadband level. DES simulation of the same grid also underpredicts PSD, particularly at low frequency outlet spectra (see bottom of Fig. 4.6), but the trends are properly captured and the amplitude offset is much lower than in URANS (about "only" 10 dB). Broadband noise is related to turbulence, i.e., vortices of different scales (and thus frequencies). Zamiri et al. [42] modified the leading edge of the centrifugal compressor diffuser vanes and obtained a reduction of recirculating flow, separation bubbles and entropy generation. The decrease of secondary flows entailed a reduction on the overall SPL in their numerical simulations.

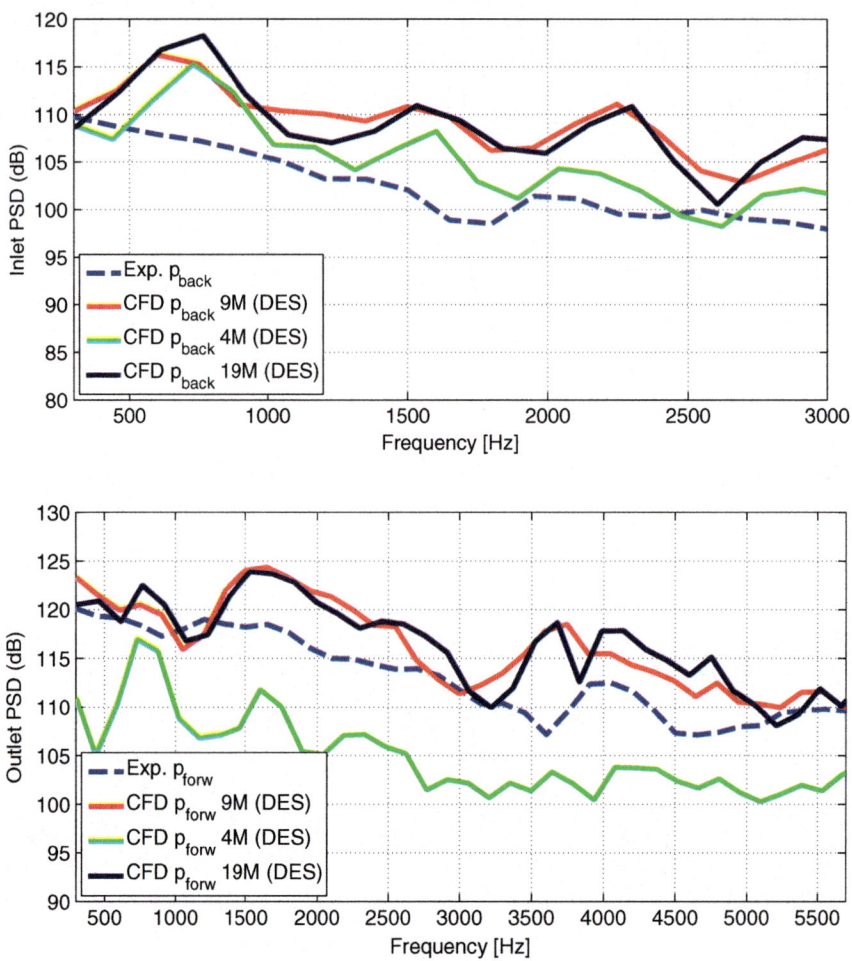

Fig. 4.6 Low frequency PSD of experimental and numerical pressure components at inlet (top) and outlet (bottom) ducts at 60 g/s for DES simulations with different meshes

DES calculations improve noise prediction over URANS simulations using the same mesh, especially regarding 4 million elements grid. Since URANS and RANS branch of DES simulations in this work use the same $k - \omega$ SST model, the only difference that DES calculations present is the amount of domain being solved with LES. Taking into account Eq. 17 in the work of Shur et al. [34], $\widetilde{f_d}$ is the parameter responsible for blending RANS and LES turbulent length scales when dealing with IDDES mode. For instance, Greschner et al. [43] used this blending function to depict regions working with RANS or LES mode in their IDDES airfoil simulation. However, a current limitation in StarCCM+ [1] is that the code does not allow one to display $\widetilde{f_d}$.

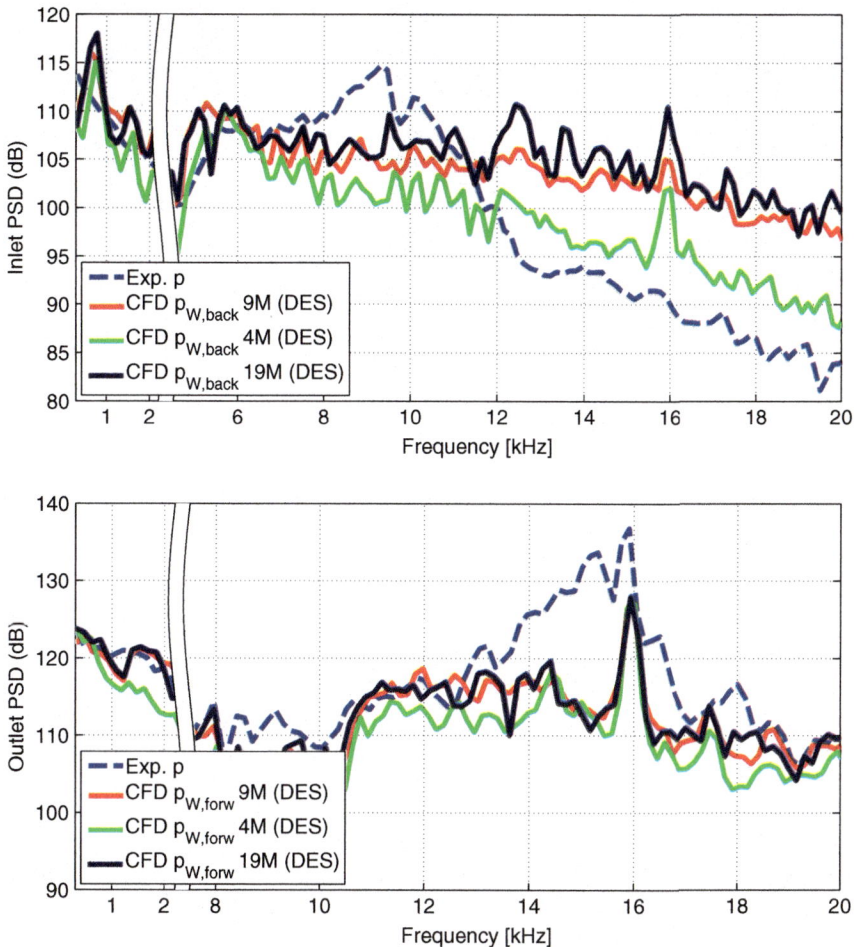

Fig. 4.7 High frequency PSD of experimental and numerical pressure components at inlet (top) and outlet (bottom) ducts at 60 g/s for DES simulations with different meshes

A workaround was found by switching the IDDES simulation to DDES. In this way, the DDES correction factor ϕ could be displayed, being the unity in RANS zone and greater than one in LES region. This approach neglects the differences in implementation that may exist between pure DDES and the DDES branch of IDDES. Moreover, WMLES branch of IDDES may increase the amount of LES-resolved cells near the walls. Values of ϕ would therefore represent just an estimation of the domain resolved in LES.

Figure 4.8 presents ϕ contours over a cross-sectional plane of the whole domain, along with mesh resolution. DDES correction factor has been time-averaged over 3 impeller revolutions. Cells presenting ϕ values above two (corresponding to LES

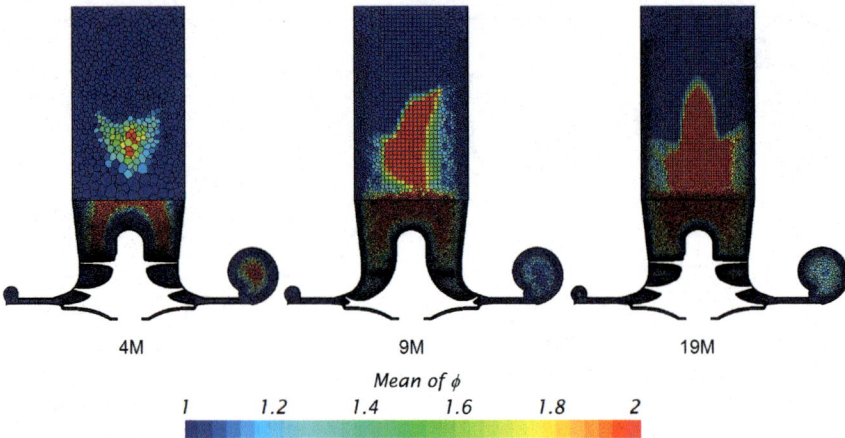

4M   9M   19M

Mean of φ

1   1.2   1.4   1.6   1.8   2

Fig. 4.8 Estimation of RANS and LES zones at cross-sectional plane for DES cases with different meshes

region) are also represented in red, because the objective of Fig. 4.8 is not obtaining the accurate value of DDES correction factor but knowing the distribution of RANS and LES modes.

Figure 4.8 shows that LES mode (colours different from dark blue) is used in the last section of the inlet duct, in the rotor and in the volute. Upstream the impeller, LES region extends to the height at which recirculation exists at this operating condition, as will be detailed in Sect. 5.4. Increment of cell amount from 9 to 19 million expands LES mode near the walls of the inlet duct. But a larger difference exists between 9 million and 4 million grid, in which only an island of LES region appears at the inlet duct. This zone may appear by the poor resolution of coarsest mesh near the interface between inlet duct and rotor and because at that axial position, backflows are reintegrated into main stream, thus increasing the amount of detached eddies.

At the volute, the mesh using 9 million elements presents coarse cells in the cross-section core, being therefore constrained to resolve this zone in RANS mode. The finest mesh presents LES region even in the core of the outlet duct (not shown here). In the rotor region, the two finest meshes resolve in LES mode the same amount of cells, whereas the grid with 4 million elements present more RANS-resolved region. Figure 4.9 depicts a detail of ϕ contours at impeller midspan, confirming the similarity between the two finest meshes and the decrease in LES-resolved region in the core of the passages when 4 million elements are used. Midspan surfaces are produced following the approach described in Sect. A.2.2.

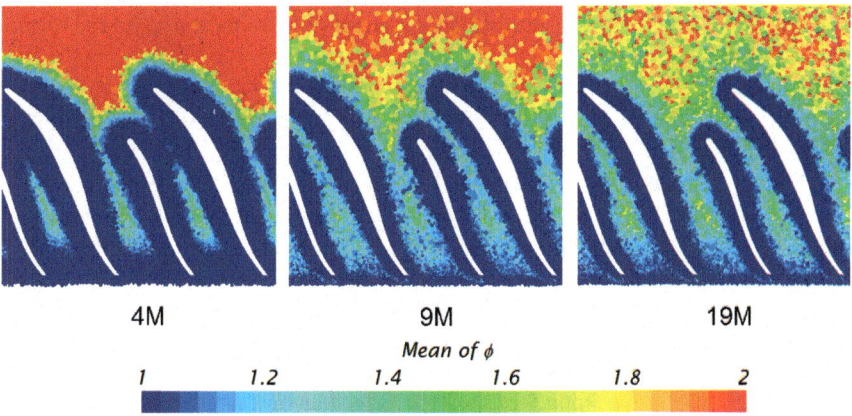

Fig. 4.9 Estimation of RANS and LES zones at impeller midspan for DES cases with different meshes

4.5 Solver

StarCCM+ [1] offers two types of solver: segregated or coupled. In the segregated solver, a pressure equation is derived from continuity and momentum equations. The governing equations are resolved in a sequential fashion, thus requiring a pressure-velocity coupling method (more details can be obtained in the book by Ferziger and Perić [44]). Conversely, coupled solver deals with Navier/Stokes equations at the same time, thus demanding more RAM memory during the simulation. Riemann solvers (see the book of Toro [45]) can be implemented into a coupled solver to deal with the inviscid part of the flow, thus improving the shock-capturing capabilities.

Traditionally, segregated solvers (also known as pressure-based solvers) have been used for incompressible flows to low-speed compressible flows, whereas coupled solvers (or density-based solvers) dealt with high-speed compressible flows. Currently, numerical improvements allow these solvers to perform properly for a wide range of mach numbers. Particularly, coupled solvers use *preconditioning* to alleviate the numerical stiffness encountered in low-speed flows. StarCCM+ [1] recommends segregated solver for aeroacoustic simulations because its lower computational cost. Nevertheless, coupled solver seems a better option for the transonic flow that appears for operating points near choke. However, preconditioning may negatively affect the propagation of low amplitude pressure disturbances at small time scales in regions with low mach numbers (see for instance Sachdev et al. [46]), such as the inlet and outlet duct. It is therefore justified to perform a comparison between DES and URANS simulations with both segregated (baseline) and coupled solvers.

Table 4.2 includes global variables (Eqs. 2.3 and 2.4) along with relative difference against experimental measurements (Eq. 2.5) and relative difference against baseline case (Eq. 4.2, in which the segregated solver is used in the baseline case) with the corresponding turbulence model for simulations performed with segregated

Table 4.2 Experimental and numerical compressor global variables for segregated and coupled solvers at 60 g/s

Case		$\Pi_{t,t}$ [-] (ϕ/ϵ_{exp}/ϵ_{base})	W_u [kJ·kg^{-1}] (ϕ/ϵ_{exp}/ϵ_{base})	η_s [%] (ϕ/ϵ_{exp}/ϵ_{base})
Exp.		2.22/–/–	121/–/–	62.2/–/–
URANS	Segr. coupled	2.21/−0.5/–	122/0.7/–	62.3/0.1/–
		2.21/−0.6/0.0	122/0.6/−0.1	62.3/0.2/0.1
DES	Segr. coupled	2.23/0.4/–	122/0.6/–	63.2/1.6/–
		2.24/0.7/0.3	122/0.2/−0.4	63.7/2.4/0.8

and coupled solver. The impact of the type of solver on compressor performance indicators is more important for DES, although relative differences between solvers do not exceed 1%.

Figures 4.10, 4.11, 4.12 and 4.13 depict PSD for URANS and DES simulations performed with segregated and coupled solvers. Pressure spectra seems insensitive to solver type, except for inlet duct PSD above 12 kHz, where amplitude decay predicted by coupled solver is more intense.

Since the studied solvers are expected to predict different flow behavior at high speed compressible flows, contours of relative Mach number, defined as

$$M_r = \frac{|\vec{v} - \vec{\omega} \wedge \vec{r}|}{a}, \tag{4.3}$$

are depicted at midspan using segregated and coupled solvers for 60 g/s mass flow rate in Fig. 4.14.

The snapshots shown at Fig. 4.14 present maximum relative Mach numbers about one in the vicinity of the leading edge for the coupled solver. The working point selected for the setup analysis is 60 g/s, which presents low mass flow rate for 160 krpm compressor speed and is thus close to surge. As stated in Sect. 4.1, this operating condition has been chosen because flow is expected to be more complicated and the model may be more sensitive to the numerical configuration. This does not hold for the solver type comparison, for which operating conditions with higher mass flow rate could have presented more differences. For instance, Fig. 4.15 advances the flow field of a case with 109 g/s, which will be studied in Chaps. 5 and 6.

Relative mach numbers close to the leading edge for 109 g/s are above the unity, which confirms the possibility of a different performance between segregated and coupled solvers. In any case, centrifugal compressor noise is reduced as working point moves away from surge line, as will be confirmed in Chap. 6.

All in all, the only difference between solvers in terms of noise prediction at 60 g/s is the inlet duct PSD decay rate (see Figs. 4.11 and 4.13). This feature is not worth the increase in computational effort that involves the selection of the coupled solver, which demands twice as much RAM memory as the segregated solver, so the latter solver will be used for the rest of the book.

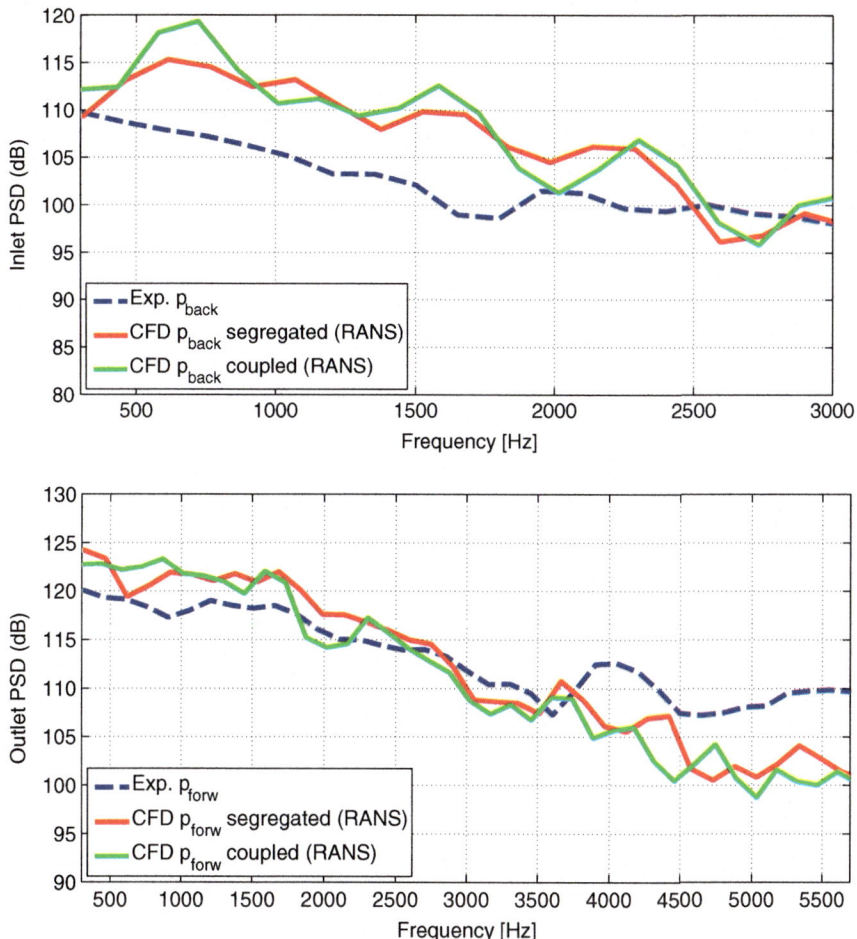

Fig. 4.10 Low frequency PSD of experimental and numerical pressure components at inlet (top) and outlet (bottom) ducts at 60 g/s for URANS simulations with different solvers

4.6 Time-Step Size

As explained in Sect. 4.2, rigid body motion simulations are transient. Definition of the appropriate time-step size, Δt, is particularly important in this work because a proper prediction of the compressor flow-induced noise is sought. For the sake of accuracy, the unsteady term is discretized using an second-order scheme. This temporal scheme is implicit, thus not setting any constraint on time-step size because it is supposed to be unconditionally stable.

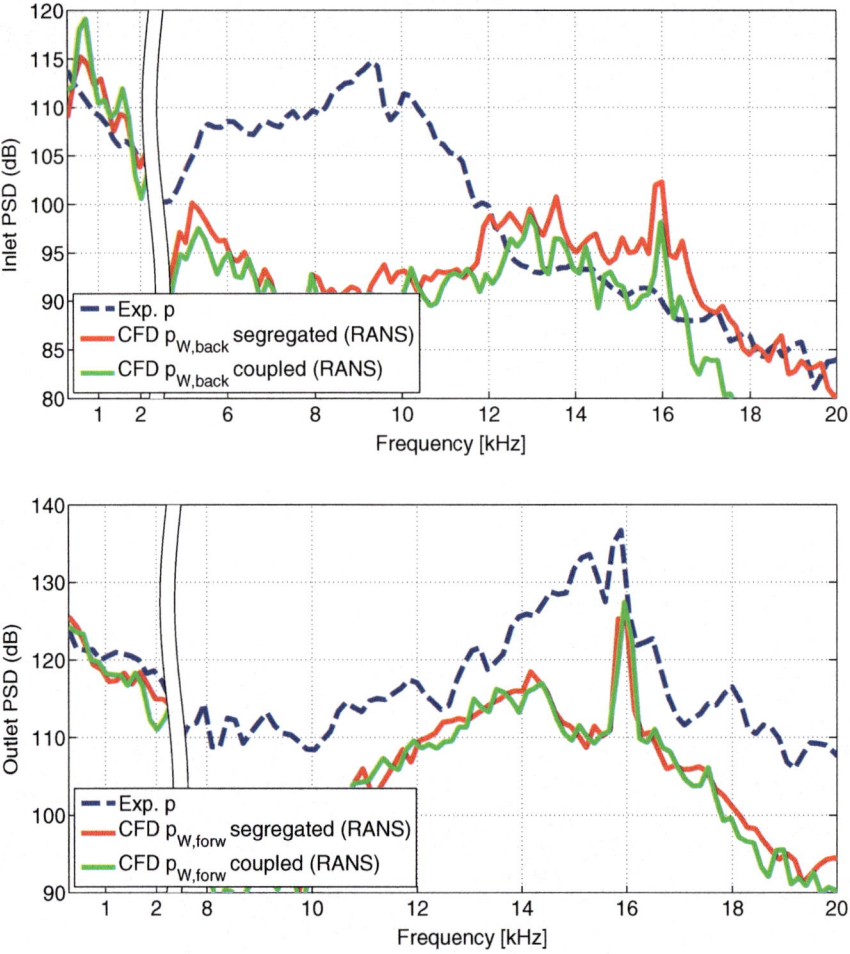

Fig. 4.11 High frequency PSD of experimental and numerical pressure components at inlet (top) and outlet (bottom) ducts at 60 g/s for URANS simulations with different solvers

Several criteria should be taken into account in order to select an adequate time-step size. First, Nyquist condition implies that simulations performed with a certain time-step size Δt are not able to resolve frequencies above

$$f_N = \frac{1}{2\Delta t}. \tag{4.4}$$

In this way, one could use the time-step increment that provides a Nyquist frequency f_N equal to the maximum human hearing frequency (about 20 kHz), which results in $\Delta t = 2.5 \times 10^{-5}$. Such a time-step size would not guarantee proper temporal resolution up to 20 kHz, because using two points per wavelength (Nyquist

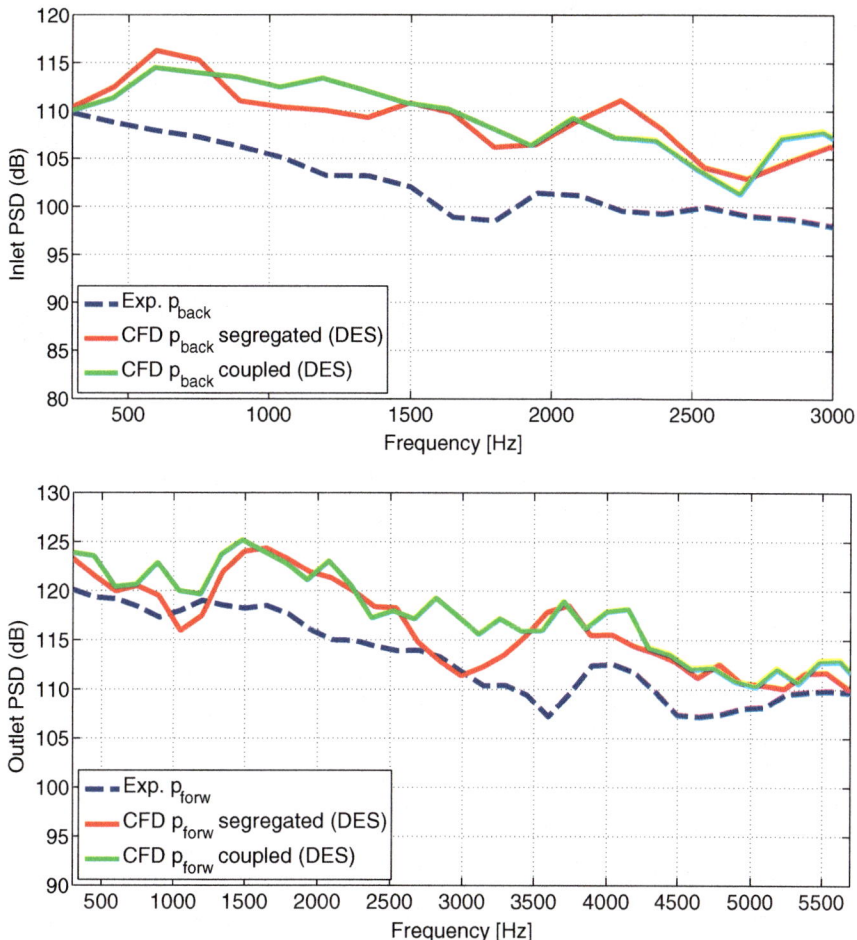

Fig. 4.12 Low frequency PSD of experimental and numerical pressure components at inlet (top) and outlet (bottom) ducts at 60 g/s for DES simulations with different solvers

criterion) may not describe the amplitude at this frequency properly and other aspects like numerical damping may affect the temporal accuracy. In fact, Mendonça [47] suggests using ten points per wavelength, which provides $\Delta t = 5 \times 10^{-6}$ to resolve 20 kHz.

For turbomachinery simulations, time-step increments are often related to the rotor speed, following a Strouhal-like approach. Temporal discretization is thus expressed in degrees rotated by the impeller per time step. For instance, each blade passing should be resolved using at least 20 time steps if an accurate prediction of rotor-stator interaction was intended [48]. Table 4.3 presents a literature survey of time-step size used in simulation of vaneless centrifugal compressors. Time-step increment is expressed in seconds, in degrees rotated by the impeller and in amount of time steps used to resolve each blade passing (BP).

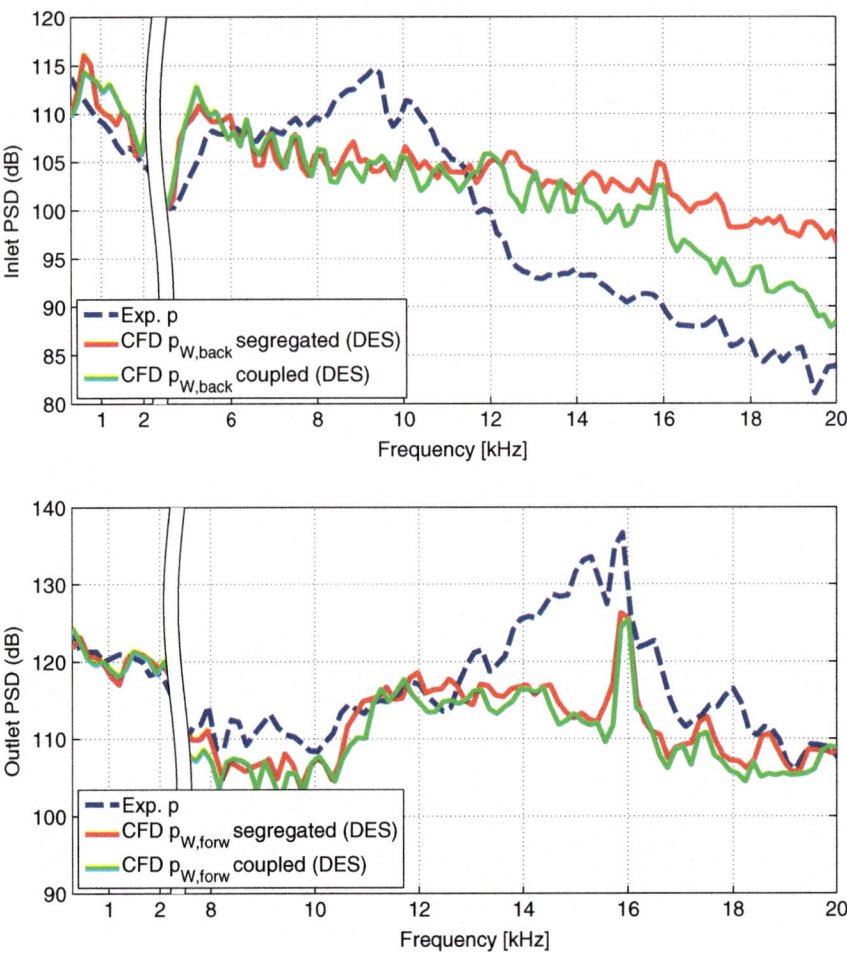

Fig. 4.13 High frequency PSD of experimental and numerical pressure components at inlet (top) and outlet (bottom) ducts at 60 g/s for DES simulations with different solvers

Analyzing Table 4.3, the time-step size used by Després et al. [49] is one order of magnitude lower than the rest of the values in the literature review. It may be due to other constrains, such as the CFL condition, i.e., the restriction in Courant number for explicit time integration schemes [44]. The rest of time-step sizes are included within one order of magnitude. The value used by Mendonça et al. [2], $\Delta t = 1°$, lies in the middle of the range, so it makes sense to test other time-step sizes around this value. In this way, simulations have been performed using time-step sizes of 0.5°, 1° (baseline), 2° and 4°. It is worth mentioning that the computational effort is practically doubled if the time-step size is divided by two, which highlights the importance of selecting a good trade-off in this setup parameter.

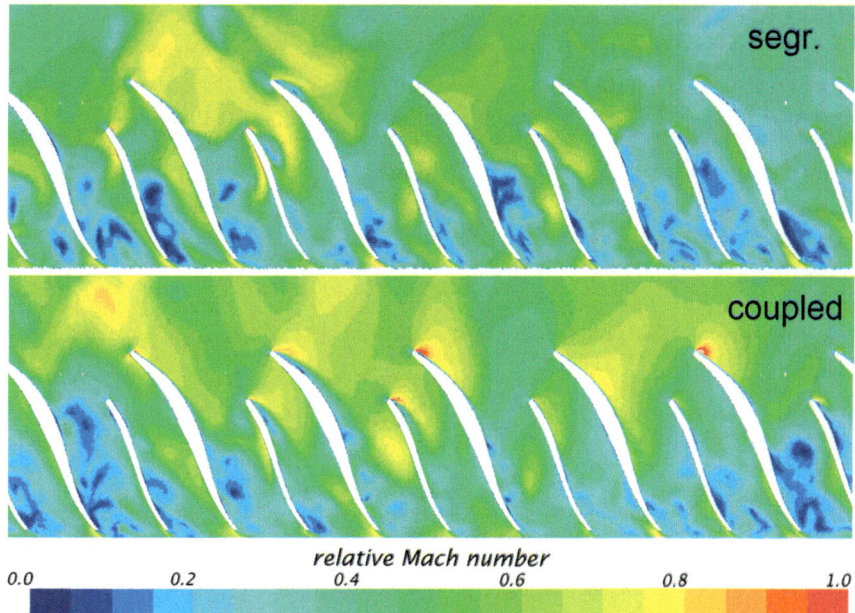

Fig. 4.14 Snapshots of relative mach number contours at impeller midspan for 60 g/s case computed with segregated (top) and coupled (bottom) solvers

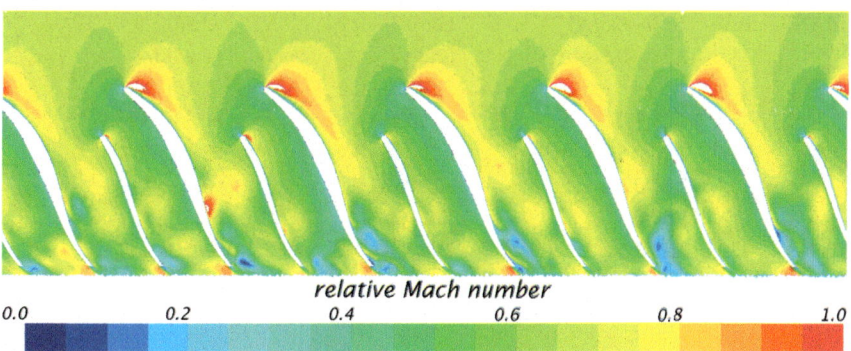

Fig. 4.15 Relative mach number contours at impeller midspan for 109 g/s case computed with segregated solver

Table 4.3 Time-step size literature survey

Paper	Δt [s]	Δt [°]	BP time steps
Després et al. [49]	1.6×10^{-8}	0.01	2500
Turunen and Larjola [50]	1×10^{-6}	0.13	198
Hellström et al. [22]	5×10^{-7}	0.19	188
Pitkänen et al. [51]	2×10^{-6}	0.22	119
Fontanesi et al. [52]	8.33×10^{-7}	0.5	60
Lee et al. [40]	$2.09/2.35 \times 10^{-6}$	1	45
Mendonça et al. [2]	7.94×10^{-7}	1	30
Guo et al. [38] Chen et al. [39]	1.13×10^{-5}	3	10
Tomita et al. [41]	$3.75/7.5 \times 10^{-6}$	3.6/7.2	10/6
Jyothishkumar et al. [23] Semlitsch et al. [9]	1.3×10^{-5}	5	7

Table 4.4 Experimental and numerical compressor global variables for time-step size sensitivity analysis at 60 g/s

Case		$\Pi_{t,t}$ [–] ($\phi/\epsilon_{exp}/\epsilon_{base}$)	W_u [kJ · kg^{-1}] ($\phi/\epsilon_{exp}/\epsilon_{base}$)	η_s [(%)] ($\phi/\epsilon_{exp}/\epsilon_{base}$)
Exp.		2.22/–/–	121/–/–	62.2/–/–
URANS	0.5°	2.21/−0.5/0.1	122/0.5/−0.2	62.5/0.5/0.4
	1°	2.21/−0.5/–	122/0.7/–	62.3/0.1/–
	2°	2.21/−0.6/0.0	122/0.5/−0.2	62.4/0.2/0.1
	4°	2.21/−0.6/−0.1	122/0.7/0.0	62.3/0.0/−0.1
DES	0.5°	2.23/0.5/0.1	122/0.5/−0.1	63.4/1.8/0.2
	1°	2.23/0.4/–	122/0.6/–	63.2/1.6/–
	2°	2.23/0.5/0.0	122/0.5/−0.1	63.3/1.8/0.1
	4°	2.23/0.5/0.1	122/0.3/−0.3	63.5/2.0/0.4

Total-to-total pressure ratio, specific work and isentropic efficiency are calculated using Eqs. 2.3 and 2.4 for the simulation varying time-step size. Table 4.4 includes these global variables along with relative difference against experimental measurements (Eq. 2.5) and relative difference against baseline case (Eq. 4.2, in which $\Delta t = 1°$ is used in the baseline case) with the corresponding turbulence model for these simulations.

Compressor performance variables are calculated with relative differences below 0.5% regardless of the time-step size used. If prediction of compressor global variables were the only objective, computational effort could be reduced by a factor of 4 from current configuration provided that simulations were performed using $\Delta t = 4°$ instead. However, the influence of time-step increment on PSD should be

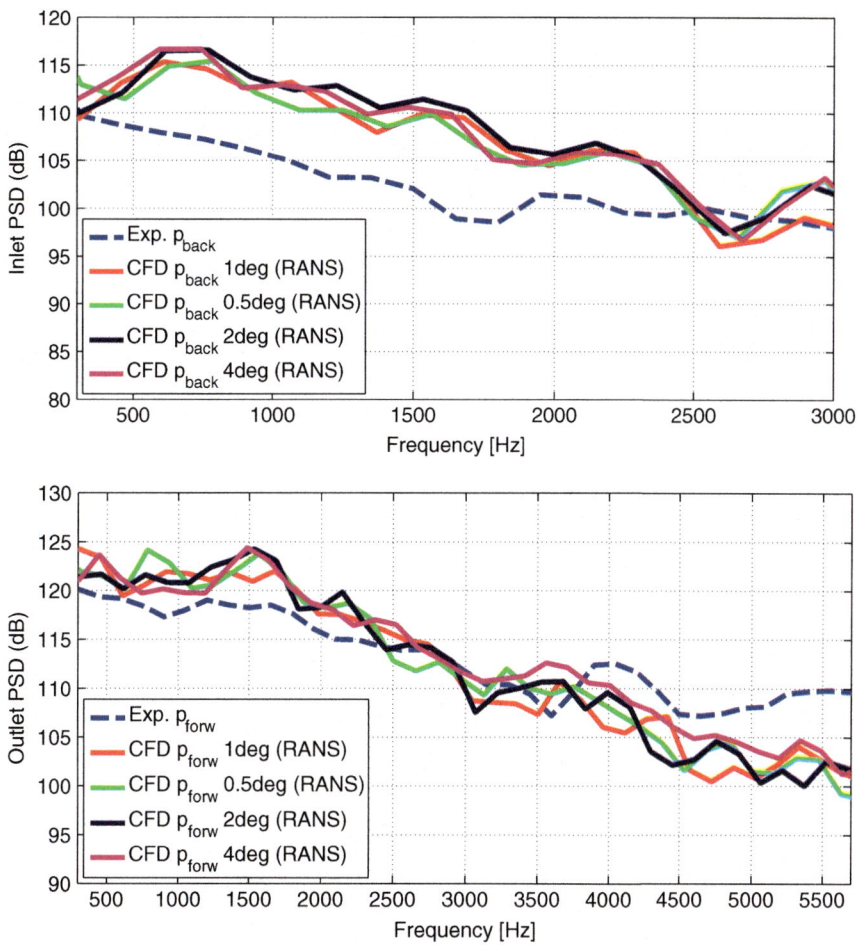

Fig. 4.16 Low frequency PSD of experimental and numerical pressure components at inlet (top) and outlet (bottom) ducts at 60 g/s for URANS simulations with different time-step sizes

also assessed. Figures 4.16, 4.17, 4.18 and 4.19 depict pressure spectra for URANS and DES simulations using different time-step sizes.

In the low frequency range, below onset of first high-order mode, no significant discrepancies are found between simulations with different time-step size (see Figs. 4.16 and 4.18). Since human hearing sensitivity peak lies in this low frequency range, computations could be performed with $\Delta t = 4°$ and prediction of overall SPL would be acceptable, However, observation of high frequency range (Figs. 4.17 and 4.19) implies that PSD are properly predicted only up to a frequency that increases with decreasing time-step size. Note that frequency upper limit in Figs. 4.17 and 4.19 has been extended from usual 20 to 30 kHz. Spectra obtained with $\Delta t = 4°$ simulation departs from the rest of PSD at about 10 kHz. When $\Delta t = 2°$ is used, blade

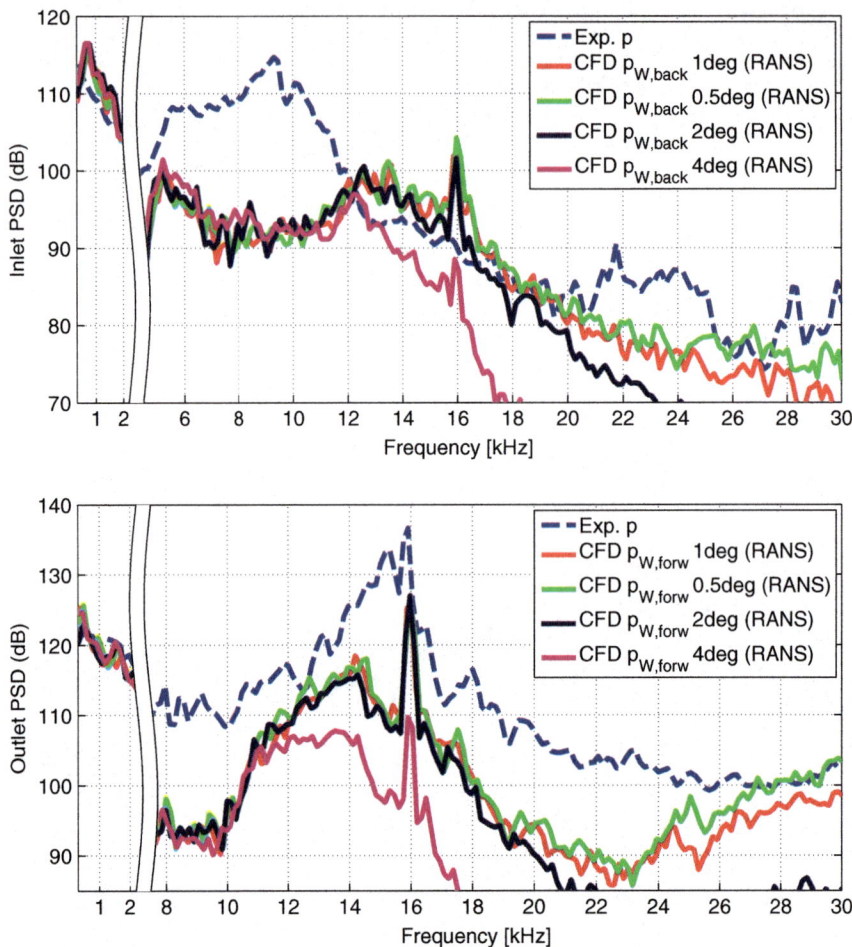

Fig. 4.17 High frequency PSD of experimental and numerical pressure components at inlet (top) and outlet (bottom) ducts at 60 g/s for URANS simulations with different time-step sizes

passing tone is properly captured but the spectra are dampened above 18 kHz. If $\Delta t = 1°$, first differences regarding PSD simulated with the smallest time-step size (0.5°) are only found above 25 kHz. Taking into account that this book is devoted to the analysis of turbocharger compressor aeroacoustics, remaining simulations will be performed with baseline time-step increment, i.e., $\Delta t = 1°$, for the sake of frequency resolution. For industrial simulations, it seems interesting, though, to increase time-step size to $\Delta t = 2°$ or even $\Delta t = 4°$ so as to reduce the computational effort.

It should be noticed that these findings may not hold for different grid resolutions, particularly for Detached Eddy Simulations. Spalart [53] estimated that a CFL of about 1 should be used in the LES region of a DES to maintain the temporal accuracy.

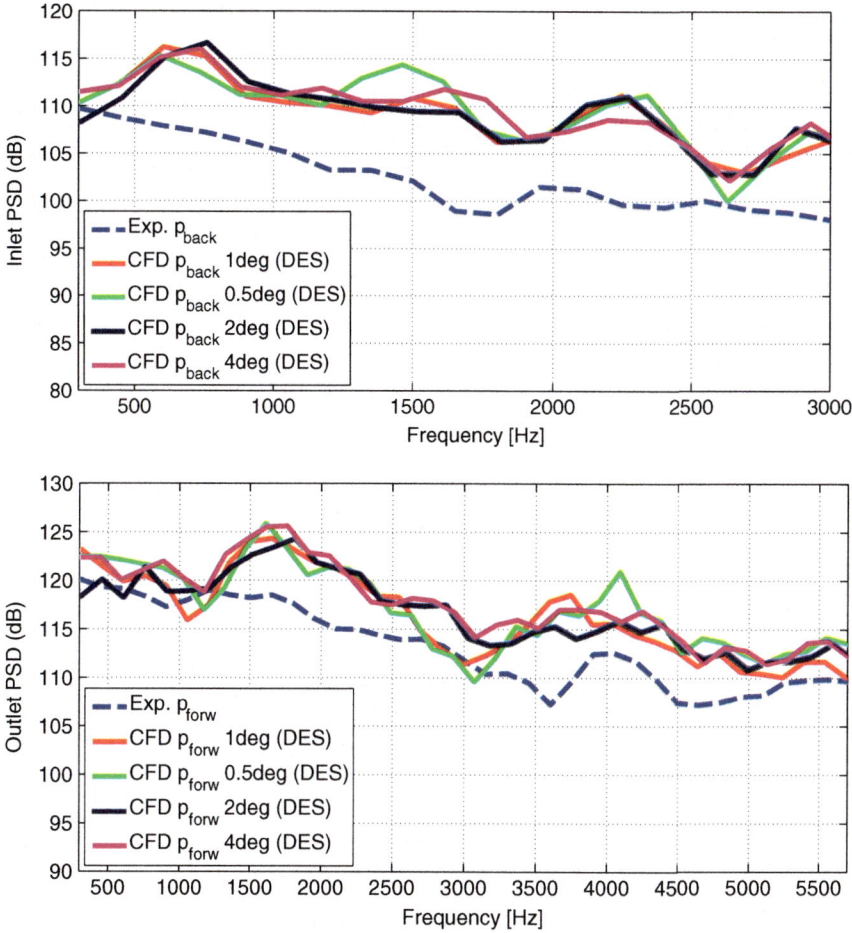

Fig. 4.18 Low frequency PSD of experimental and numerical pressure components at inlet (top) and outlet (bottom) ducts at 60 g/s for DES simulations with different time-step sizes

Figure 4.20 presents snapshots of CFL contours at a cross-sectional plane for DES simulations using $\Delta t = 1°$ and $\Delta t = 4°$. Most of LES domain (see middle picture in Fig. 4.8) is calculated with a CFL number less than one for $\Delta t = 1°$, whereas the rotor region for $\Delta t = 4°$ presents Courant number exceeding the unity.

Courant number contours at midspan for cases with $\Delta t = 1°$ and $\Delta t = 4°$ are depicted in Fig. 4.21.

For $\Delta t = 1°$, stagnation point at blades leading edge is calculated with a Courant number above one, although Fig. 4.9 indicates that the vicinity of the blades is in RANS mode. Conversely, case with $\Delta t = 4°$ presents high Courant numbers upstream the blades and at the core of the passages, which belong to the LES zone

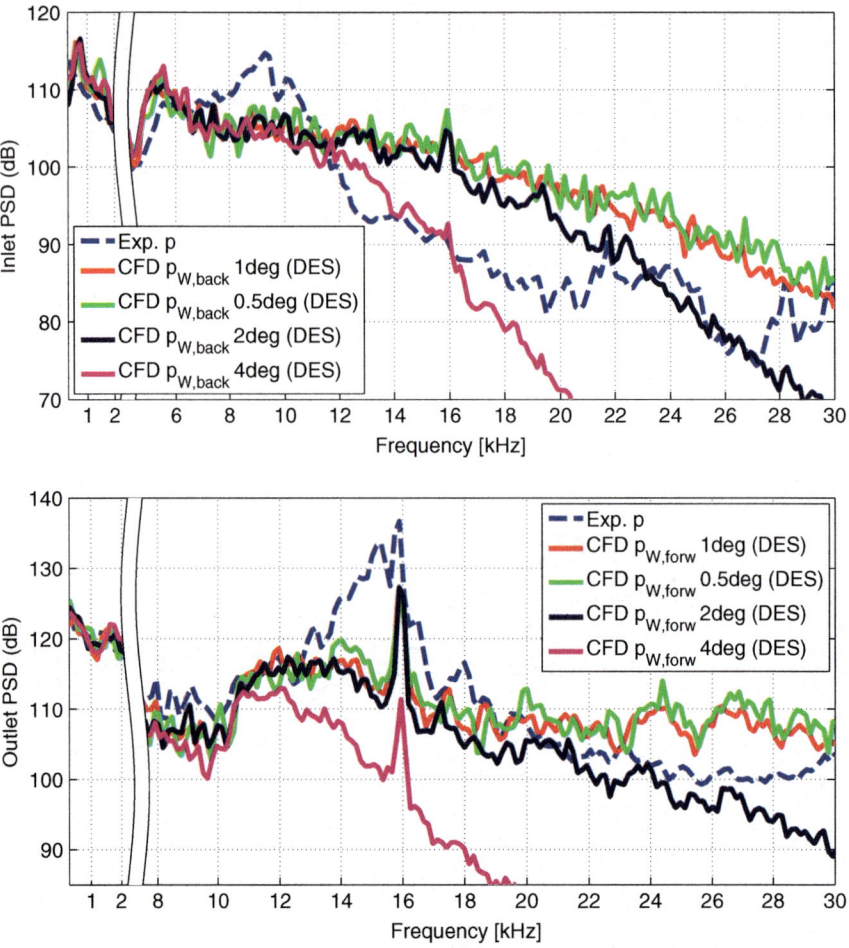

Fig. 4.19 High frequency PSD of experimental and numerical pressure components at inlet (top) and outlet (bottom) ducts at 60 g/s for DES simulations with different time-step sizes

(see Fig. 4.9). According to Spalart [53], these large Courant numbers may deteriorate temporal accuracy in the case with $\Delta t = 4°$, as confirmed by Fig. 4.19.

To conclude, it should be noted that Cummings et al. [54] provide valuable information about the factors that have an influence on the temporal accuracy of unsteady simulations, proposing rules of thumb for numerical options such as time-step size.

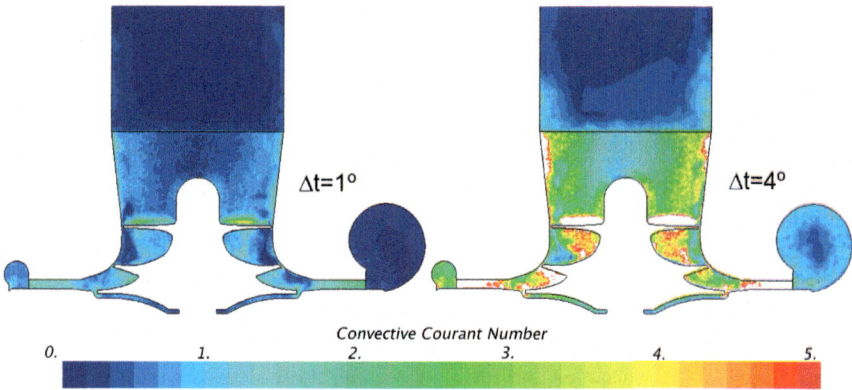

Fig. 4.20 Snapshot of CFL contours at a cross-sectional plane for DES simulations using $\Delta t = 1°$ (left) and $\Delta t = 4°$ (right)

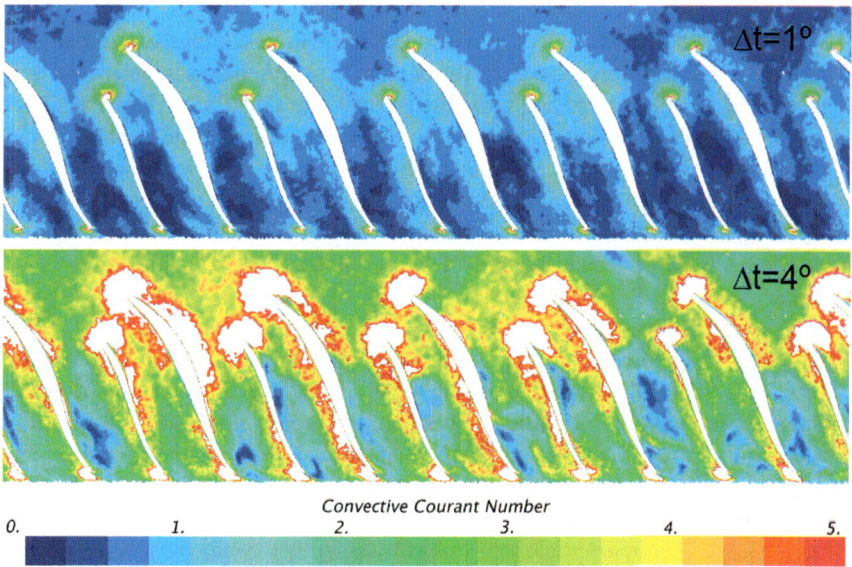

Fig. 4.21 Snapshot of CFL contours at impeller midspan for DES simulations using $\Delta t = 1°$ (top) and $\Delta t = 4°$ (bottom)

4.7 Conclusions

In this chapter, the numerical configuration of the model has been discussed. Unsteady rigid body motion simulations are required to resolve the aeroacoustics of centrifugal compressors. DES turbulence model is expected to present a better broadband noise resolution than URANS simulations, due to its eddy-resolving capabilities. In order to test this feature, both turbulence approaches will be used to compare their predicted spectra.

Besides, grid spacing and time-step size used by Mendonça et al. [2] have proved to be right choices, although now the sensitivity of compressor performance parameters and noise production to these options is known. This information can be used to select appropriate setup trade-off if simulations workaround time is required to be reduced (for industrial cases, for instance).

When the mesh is refined from 4 million elements to 9 million cells, y^+ values move from the buffer region to the viscous sublayer in almost all the domain. Moreover, LES branch of IDDES is extended, particularly at the zone of the inlet duct where recirculating flow appears. These two effects have an impact on both compressor performance variables and PSD. Increasing mesh density until reaching 19 million cells grid improves viscous sublayer resolution and slightly increases LES-resolved region, although no significant modifications in global variables and noise prediction exist. Apart from a slight better prediction of compressor noise generation, DES proves to be less sensitive to mesh size than URANS.

Segregated and coupled solver provide almost identical compressor global variables in URANS simulations. In DES, differences between solvers increase, but they do not exceed 1%. In terms of pressure spectra, only inlet duct PSD at high frequency differs, because coupled solver is found to predict a steeper amplitude decay above 12 kHz (for both URANS and DES turbulence models). Therefore, for 60 g/s working point the increase in RAM memory required by coupled solver is not worth it, and the segregated solver is therefore selected. At higher mass flow rates, for which relative Mach number locally exceeds the unity, the discrepancies between both solvers may be higher.

Finally, simulations using time-step sizes between $\Delta t = 0.5°$ and $\Delta t = 4°$ have been investigated. Compressor global variables can be predicted with relative difference less than 1% with any time-step increment. However, as the time-step size is increased, the maximum frequency at which the predicted spectra is accurate is reduced. In this work, the value $\Delta t = 1°$ is selected so as to maintain accuracy, but if computational effort is an issue, even $\Delta t = 4°$ cases deliver PSD predictions that are adequate in the low frequency spectra.

Having defined an appropriate setup to study centrifugal compressor flow-induced acoustics, the evolution of acoustic signature and aeroacoustic phenomena with operating conditions will be evaluated in the remainder of the book. Before dealing with transient features, a description of the changes experienced by mean flow when mass flow rate is reduced is presented in Chap. 5.

References

1. STAR-CCM+. Release version 9.02.005. CD-adapco (2014), http://www.cd-adapco.com
2. F. Mendonça, O. Baris, G. Capon, Simulation of radial compressor aeroacoustics using CFD. In *Proceedings of ASME Turbo Expo 2012. GT2012-70028*, ASME, 2012, pp. 1823–1832 (2012). https://doi.org/10.1115/GT2012-70028
3. X.Q. Zheng, J. Huenteler, M.Y. Yang, Y.J. Zhang, T. Bamba, Influence of the volute on the flow in a centrifugal compressor of a high-pressure ratio turbocharger. Proc. Inst. Mech. Eng. Part A J. Power Energy **224**(8), 1157–1169 (2010). https://doi.org/10.1243/09576509JPE968, http://pia.sagepub.com/content/224/8/1157.full.pdf+html
4. J. Benajes, J. Galindo, P. Fajardo, R. Navarro, Development of a segregated compressible flow solver for turbomachinery simulations. J. Appl. Fluid Mech. **7**(4), 673–682 (2014)
5. K. Hillewaert, R. Van den Braembussche, Numerical simulation of impeller-volute interaction in centrifugal compressors. J. Turbomach. **121**, 603 (1999)
6. Z. Liu, D. Hill, Issues surrounding multiple frames of reference models for turbo compressor applications. In: *Fifteenth International Compressor Engineering Conference*, Purdue University, 2000
7. J. Galindo, P. Fajardo, R. Navarro, L.M. García-Cuevas, Characterization of a radial turbocharger turbine in pulsating flow by means of CFD and its application to engine modeling. Appl. Energy **103**, 116–127 (2013). https://doi.org/10.1016/j.apenergy.2012.09.013
8. M. Padzillah, S. Rajoo, R. Martinez-Botas, Influence of speed and frequency towards the automotive turbocharger turbine performance under pulsating flow conditions. Energy Convers. Manage. **80**(0), 416–428 (2014). ISSN 0196-8904. https://doi.org/10.1016/j.enconman.2014.01.047
9. B. Semlitsch, V. JyothishKumar, M. Mihaescu, L. Fuchs, E. Gutmark, M. Gancedo, Numerical flow analysis of a centrifugal compressor with ported and without ported shroud. SAE Technical Paper 2014-01-1655 (2014). https://doi.org/10.4271/2014-01-1655
10. Y. Bousquet, X. Carbonneau, G. Dufour, N. Binder, I. Trebinjac, Analysis of the unsteady flow field in a centrifugal compressor from peak efficiency to near stall with full-annulus simulations. Int. J. Rotating Mach. **2014**, 11 (2014). https://doi.org/10.1155/2014/729629
11. K. Jiao, H. Sun, X. Li, H. Wu, E. Krivitzky, T. Schram, L. Larosiliere, Numerical simulation of air flow through turbocharger compressors with dual volute design. Appl. Energy **86**(11), 2494–2506 (2009). ISSN 0306-2619. https://doi.org/10.1016/j.apenergy.2009.02.019
12. H.K. Versteeg, W. Malalasekera, *An Introduction to Computational Fluid Dynamics: The Finite volume method*, 2nd edn. (Pearson Education Limited, Harlow, 2007)
13. M. Gageik, I. Klioutchnikov, H. Olivier, Comprehensive mesh study for a direct numerical simulation of the transonic flow at $Re_c = 500,000$ around a NACA 0012 airfoil. Comput. Fluids **122**, 153–164 (2015). https://doi.org/10.1016/j.compfluid.2015.08.030
14. S. Hoyas, J. Jiménez, Scaling of the velocity fluctuations in turbulent channels up to $Re_\tau = 2003$. Phys. Fluids (1994–Present) **18**(1), 4 (2006). https://doi.org/10.1063/1.2162185
15. D.C. Wilcox, *Turbulence Modeling for CFD (Hardcover)*, 3rd edn. (La Cañada, California, DCW Industries Inc., 2006)
16. J. Smagorinsky, General circulation experiments with the primitive equations. Mon. Weather Rev. **91**(3), 99–164 (1963)
17. J.W. Deardorff, A numerical study of three-dimensional turbulent channel flow at large Reynolds numbers. J. Fluid Mech. **41**(02), 453–480 (1970)
18. C.A. Wagner, T. Hüttl, P. Sagaut (eds.), *Large-Eddy Simulation for Acoustics* (Cambridge University Press, Cambridge, 2007)
19. G. Dufour, N. Gourdain, F. Duchaine, O. Vermorel, L. Gicquel, J.F. Boussuge, T. Poinsot, Large eddy simulation applications. In *Numerical Investigations in Turbomachinery: A State of the Art*. VKI Lecture Series (2009)
20. A. Karim, K. Miazgowicz, B. Lizotte, A. Zouani, Computational aero-acoustics simulation of compressor whoosh noise in automotive turbochargers. SAE Technical Paper 2013-01-1880 (2013). https://doi.org/10.4271/2013-01-1880

21. F. Hellström, E. Guillou, M. Gancedo, R. DiMicco, A. Mohamed, E.Gutmark, L. Fuchs, Stall development in a ported shroud compressor using PIV measurements and large eddy simulation. Technical Report. SAE Technical Paper 2010-01-0184 (2010). https://doi.org/10.4271/2010-01-0184

22. F. Hellstrom, E. Gutmark, L. Fuchs, Large eddy simulation of the unsteady flow in a radial compressor operating near surge. J. Turbomach. **134**(5), 10 (2012). https://doi.org/10.1115/1.4003816

23. V. Jyothishkumar, M. Mihaescu, B. Semlitsch, L. Fuchs, Numerical flow analysis in centrifugal compressor near surge condition. In *Fluid Dynamics and Co-located Conferences*, American Institute of Aeronautics and Astronautics, 2013, p. 13. https://doi.org/10.2514/6.2013-2730

24. R. Aghaei, A.M. Tousi, A. Tourani, Comparison of turbulence methods in CFD analysis of compressible flows in radial turbomachines. Aircr. Eng. Aerosp. Technol. **80**(6), 657–665 (2008). https://doi.org/10.1108/00022660810911608

25. O. Borm, B. Balassa, H.-P. Kau, Comparison of different numerical approaches at the centrifugal compressor RADIVER. In *20th ISABE Conference. ISABE-2011-1242*, International Society for Airbreathing Engines, 2011

26. L. Mangani, E. Casartelli, S. Mauri, Assessment of various turbulence models in a high pressure ratio centrifugal compressor with an object oriented CFD code. J. Turbomach. **134**(6), 10 (2012). https://doi.org/10.1115/1.4006310

27. F.R. Menter, Two-equation eddy-viscosity turbulence models for engineering applications. AIAA J. **32**(8), 1598–1605 (1994)

28. F.R. Menter, R. Langtry, T. Hansen, CFD simulation of turbomachinery flows-verification, validation and modeling. In *European Congress on Computational Methods in Applied Sciences and Engineering*, ECCOMAS, 2004

29. C. Robinson, M. Casey, B. Hutchinson, R. Steed, Impeller-diffuser interaction in centrifugal compressors. In *Proceedings of ASME Turbo Expo 2012. GT2012-69151*, ASME, 2012

30. P.E. Smirnov, T. Hansen, F.R. Menter, Numerical simulation of turbulent flows in centrifugal compressor stages With different radial gaps. In *Proceedings of GT2007. GT2007-27376*, ASME, 2007. https://doi.org/10.1115/GT2007-27376

31. R.N. Pinto, A. Afzal, L.V. D'Souza, Z. Ansari, A.D. Mohammed Samee, Computational fluid dynamics in turbomachinery: a review of state of the art. Arch. Comput. Methods Eng. **24**(3), 467–479. ISSN 1886-1784. https://doi.org/10.1007/s11831-016-9175-2

32. P.G. Tucker, Computation of unsteady turbomachinery flows: part 2-LES and hybrids. Prog. Aerosp. Sci. **47**(7), 546–569 (2011). https://doi.org/10.1016/j.paerosci.2011.07.002

33. Detached-eddy simulations past a circular cylinder. Flow Turbul. Combust. **63**(1–4), 293–313 (2000)

34. M.L. Shur, P.R. Spalart, M.K. Strelets, A.K. Travin, A hybrid RANS-LES approach with delayed-DES and wall-modelled LES capabilities. Int. J. Heat Fluid Flow **29**(6), 1638–1649 (2008). https://doi.org/10.1016/j.ijheatfluidflow.2008.07.001

35. A.K. Travin, M.L. Shur, P.R. Spalart, M.K. Strelets, Improvement of delayed detached-eddy simulation for LES with wall modelling. In *European Conference on Computational Fluid Dynamics (ECCOMAS CFD 2006)*, 2006

36. C. Mockett, *A comprehensive study of detached-eddy simulation*. Ph.D. thesis, Technische Universität Berlin, 2009

37. R. van Rennings, K. Shi, S. Fu, F. Thiele. Delayed-detached-eddy simulation of near-stall axial compressor flow with varying passage numbers. In *Progress in Hybrid RANS-LES Modelling*, ed. by S. Fu, W. Haase, S.-H. Peng, D. Schwamborn. Notes on Numerical Fluid Mechanics and Multidisciplinary Design, vol. 117 (Springer, Berlin Heidelberg, 2012), pp. 439–448. ISBN 978-3-642-31817-7. https://doi.org/10.1007/978-3-642-31818-4_38

38. Q. Guo, H. Chen, X.C. Zhu, Z.H. Du, Y. Zhao, Numerical simulations of stall inside a centrifugal compressor. Proc. Inst. Mech. Eng. Part A J. Power Energy **221**(5), 683–693 (2007). https://doi.org/10.1243/09576509JPE417

39. H. Chen, S. Guo, X.C. Zhu, Z.H. Du, S. Zhao, Numerical simulations of onset of volute stall inside a centrifugal compressor. In *Proceedings of ASME Turbo Expo 2008: Power for Land, Sea and Air*, ASME, 2008

40. Y. Lee, D. Lee, Y. So, D. Chung, Control of airflow noise from diesel engine turbocharger. SAE Technical Paper 2011-01-0933 (2011). https://doi.org/10.4271/2011-01-0933
41. I. Tomita, S. Ibaraki, M. Furukawa, K. Yamada, The effect of tip leakage vortex for operating range enhancement of centrifugal compressor. In *Proceedings of ASME Turbo Expo 2012. GT2012-68947*, ASME, 2012
42. A. Zamiri, B.J. Lee, J.T. Chung, Numerical investigation of the inclined leading edge oiffuser vane effects on the flow unsteadiness and noise characteristics in a transonic centrifugal compressor. In *ASME Turbo Expo 2017: Turbomachinery Technical Conference and Exposition GT2017-65117 2017*, p. 15. https://doi.org/10.1115/GT2017-65117
43. B. Greschner, J. Grilliat, M.C. Jacob, F. Thiele, Measurementsand wall modeled LES simulation of trailing edge noise caused by a turbulent boundary layer. Int.J. Aeroacoustics **9**(3), 329–356 (2010). https://doi.org/10.1260/1475-472X.9.3.329
44. Ferziger, J.H. Peric, M, *Computational Methods for Fluid Dynamics*, 3rd rev. (Springer, Berlin, 2002)
45. E. Toro, *Riemann Solvers and Numerical Methods for Fluid Dynamics: A Practical Introduction*, 2nd edn. (Springer, Berlin, 1999)
46. J. Sachdev, A. Hosangadi, V. Sankaran, Improved flux formulations for unsteady low mach number flows. In *Fluid Dynamics and Co-located Conferences*, American Institute of Aeronautics and Astronautics, June 2012. https://doi.org/10.2514/6.2012-3067
47. F. Mendonça, *Industrial Aeroacoustics Analyses*. ed. by C.A. Wagner, T. Hüttl, P. Sagaut. In *Large-Eddy Simulation for Acoustics* (Cambridge University Press, Cambridge, 2007), chap. 6.8, pp. 356–377
48. ANSYS Inc., *ANSYS FLUENT 12.0 User's Guide* (ANSYS Inc., Canonsburg, 2009)
49. G. Després, G.N. Boum, F. Leboeuf, D. Chalet, P. Chesse, A. Lefebvre, Simulation of near surge instabilities onset in a turbocharger compressor. Proc. Inst. Mech. Eng. Part A J. Power Energy **227**(6), 665–673 (2013). https://doi.org/10.1177/0957650913495537
50. T. Turunen-Saaresti, J. Larjola, Measured and calculated unsteady pressure field in a vaneless diffuser of a centrifugal compressor. ed. by K.C. Hall, R.E. Kielb, J.P. Thomas. In *Unsteady Aerodynamics, Aeroacoustics and Aeroelasticity of Turbomachines* (Springer, Dordrecht, 2006), pp. 493–503
51. H. Pitkänen, H. Esa, P. Sallinen, J. Larjola, H. Heiska, T. Siikonen, Time-accurate CFD analysis of a centrifugal compressor. In *Fourth International Symposium on Experimental and Computational Aerodynamics of Internal Flows, Dresden August* vol. 31, 1999 pp. 130–139
52. S. Fontanesi, S. Paltrinieri, G. Cantore, CFD analysis of the acoustic behavior of a centrifugal compressor for high performance engine application. Energy Procedia **45**(0) (2014). *ATI 2013—68th Conference of the Italian Thermal Machines Engineering Association*, pp. 759–768. https://doi.org/10.1016/j.egypro.2014.01.081
53. P.R. Spalart, Young-Person's guide to detached-eddy simulation grids. Technical Report, NASA (Langley Research Center), 2001
54. R.M. Cummings, S.A. Morton, D.R. McDaniel, Experiences in accurately predicting time-dependent flows. Prog. Aerosp. Sci. **44**(4), 241–257 (2008). ISSN 0376-0421. https://doi.org/10.1016/j.paerosci.2008.01.001

Chapter 5
Compressor Mean Flow Field at Near-Stall Conditions

5.1 Introduction

In Chap. 4 the different setup possibilities have been explored, defining an ultimate numerical configuration that shall be used to study aeroacoustic phenomena at three different working points. However, it seems appropriate to first described mean flow behavior at the studied operating conditions, which range from best efficiency point (BEP) to near surge conditions at 160 krpm (the surge side of this isospeed). To do so, time-averaged values of fluid variables will be displayed in the different regions of the compressor in this chapter. Section A.3.2 explains the approach followed to perform the time-averaging and Sect. A.2 describes the postprocessing surfaces that are used throughout this chapter. Scales of contours an vectors size are the same for each set of pictures to allow a proper comparison between operating conditions. Since a qualitative description of the mean flow is intended, pictures will only represent DES results to avoid repetition.

First, a literature review is performed in Sect. 5.2. Then, evolution of compressor performance variables with mass flow rate is included in Sect. 5.3. After that, mean flow behavior at different operating conditions is described, following a streamwise order. Section 5.4 is devoted to the inducer flow, Sect. 5.5 covers the evolution of flow at the passages, which eventually reaches the diffuser, volute and outlet duct, included in Sect. 5.6.

5.2 Literature Review

Jyothishkumar et al. [1] analyzed the unsteady flow of a ported shroud centrifugal compressor at design and near-surge conditions using LES. Frequency decomposition of tangential velocity traces obtained at several diffuser probes was performed. A tone at 50% of rotational speed was observed at the operating point with less mass flow rate, indicating the existence of rotating stall at near-surge conditions.

© Springer International Publishing AG 2018
R. Navarro García, *Predicting Flow-Induced Acoustics at Near-Stall Conditions in an Automotive Turbocharger Compressor*, Springer Theses, https://doi.org/10.1007/978-3-319-72248-1_5

Pressure and velocity profiles at impeller inlet and diffuser outlet are quite smooth and axisymmetric at design operating conditions, whereas the existence of backflows near surge deteriorates these profiles.

Després et al. [2] analyzed the unsteady flow inside a turbocharger compressor operating near surge using CFD. In this working point, inlet recirculation near impeller shroud extended one diameter upstream inducer plane. Flow at the diffuser was not axisymmetric. For instance, circumferential profile of pressure presented a marked decrease at volute tongue position. Recirculating regions were observed at the diffuser, at both hub and shroud faces. No rotating stall was appreciated at this operating point.

Guo et al. [3] performed a numerical simulation of stall flow phenomenon inside a turbocharger centrifugal compressor with vaneless diffuser. The stage performance curve was obtained using steady and unsteady calculations and compared with experimental results. The unsteady flow simulation was performed at reduced mass flow rates (compressor stall region) with a plenum chamber model added at the compressor housing outlet. Time-varying flow fields in the diffuser and in the volute indicate that there is a circumferential standing wave inside the volute. It is proposed that the volute behaves like a tuned pipe resonator and this dictates the stall frequency.

Chen et al. [4] conducted a study using CFD to analyze the flow at the vaneless diffuser and volute of a centrifugal compressor. They showed that, only at the peak pressure ratio point, the circumferentially-averaged radial velocity spanwise distribution at the volute inlet has an inflection point, and the distribution meets the requirement of the Fjørtoft instability theorem. Moreover, impeller total pressure rise curve has a flat region in the middle before the stage stalls. In the volute discharge section, the vortex core at the cross sectional center breaks down when the stage stalls. The diffuser is the first component to stall, thus increasing the pressure rise across the volute and finally leading it to stagnate too.

5.3   Compressor Global Variables

Compressor global variables obtained with both turbulence models are first compared against experimental measurements. 3 operating conditions at 1.4, 1.8 and 2.5 times surge mass flow at 160 krpm have been simulated, corresponding to mass flow rates of 60, 77 and 109 g/s. Total-to-total pressure ratio, specific work and isentropic efficiency are calculated using Eqs. 2.3–2.4. Table 5.1 includes these overall variables for all operating conditions, along with relative difference against experimental measurements for numerical cases, previously defined in Eq. 2.5.

Numerical simulations predict compressor performance variables in agreement with experimental measurements, the error being no greater than 3%. In contrast to what may be expected, operating conditions with higher mass flow rate present greater error, despite flow is expected to be more regular and thus easier to predict by means of CFD.

Table 5.1 Experimental and numerical compressor global variables for cases at 60, 77 and 109 g/s

Case		$\Pi_{t,t}$ $(-)$ (ϕ/ϵ_{exp})	W_u (kJ · kg^{-1}) (ϕ/ϵ_{exp})	η_s (%) (ϕ/ϵ_{exp})
Exp	60	2.22/–	121/–	62.2/–
	77	2.24/–	112/–	67.9/–
	109	2.19/–	101/–	73.0/–
URANS	60	2.21/−0.5	122/0.7	62.3/0.1
	77	2.22/−0.8	112/−0.6	67.5/−0.5
	109	2.14/−2.3	99/−2.1	72.0/−1.3
DES	60	2.23/0.4	122/0.6	63.2/1.6
	77	2.22/−0.9	111/−0.8	67.6/−0.4
	109	2.13/−2.8	99/−2.6	71.8/−1.5

Figure 5.1 shows the evolution of pressure ratio, specific work and isentropic efficiency with corrected mass flow. It should be noted that experimental iso-speed line consists of 6 points, corresponding 3 of them to the simulated operating conditions.

Specific work and isentropic efficiency curves are well predicted by numerical simulations, with no great differences between URANS and DES turbulence approaches. Pressure ratio predicted by URANS simulations presents a maximum in the same mass flow rate as in experimental measurements, i.e., 77 g/s. Conversely, DES does not provide a pressure ratio turning point in this range of mass flow rates.

In terms of compressor global variables, URANS simulations show better agreement with experimental measurements than DES, particularly in pressure ratio evolution. However, differences between numerical and experimental global variables are driven by a combination of numerical errors and model uncertainty [5], and it should be noted that phenomena such as shaft motion and heat transfer have not been modeled. Therefore, performance of turbulence models should not be assessed by prediction of overall variables alone. In Sect. 6.3, spectra predicted by URANS and DES simulations at these operating conditions will be compared.

5.4 Inlet Duct and Inducer

Figure 5.2 shows the time-averaged inlet flow at a plane section for each operating point. The postprocessing surface ends at the inducer plane. Line integral convolution (LIC) [6] blended with contours of temperature is plotted along with in-plane velocity vectors. A black solid line indicates a change of sign in axial velocity. Even the case with higher mass flow (109 g/s) presents backflows, but they are confined to the impeller periphery, as will be seen later in this chapter. At 77 g/s, recirculating flow surpasses the inducer plane, but it does not even reach the impeller's eye. At 60 g/s, backflow region is thicker and extends 1 diameter upstream the impeller's

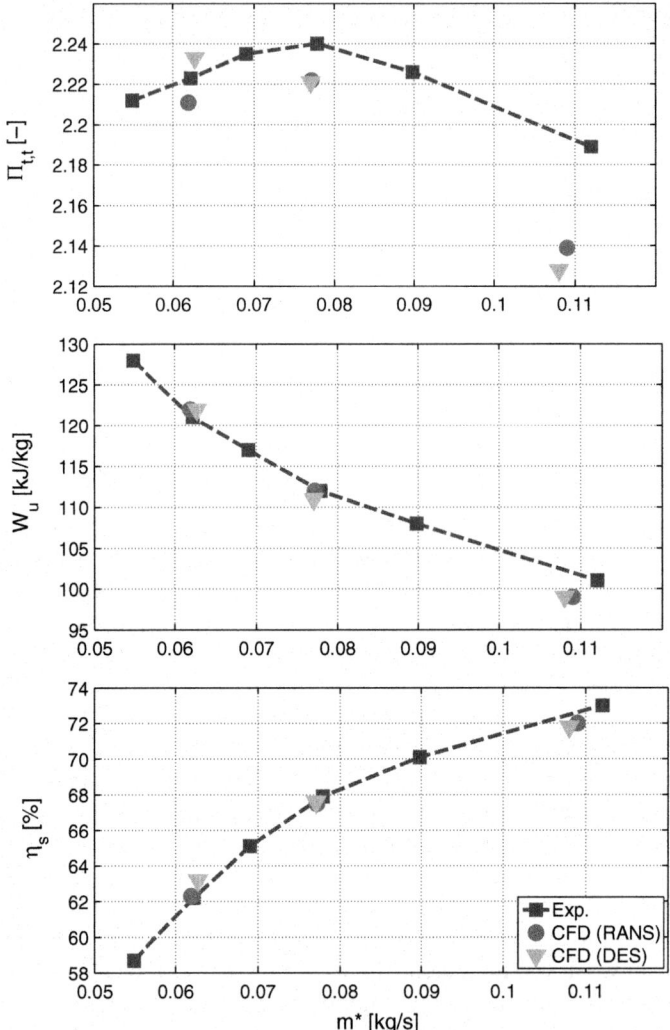

Fig. 5.1 Compressor global variables: pressure ratio (top), specific work (middle) and isentropic efficiency (bottom) versus corrected mass flow rate

eye. This recirculating flow is hotter than the incoming flow because it was partially compressed. Being a convective scalar quantity, temperature field is a good trace of backflow pattern and subsequent reintegration to the main flow.

The use of the increase of temperature at the inlet duct wall as an indicator of the onset of compressor stall and surge has been investigated by several researchers. Andersen et al. [7] showed that inlet temperature rise is detected by sensors close to the impeller even for operating conditions far from surge. An axial array of temperature probes was used, locating each sensor at a radial distance of half the pipe

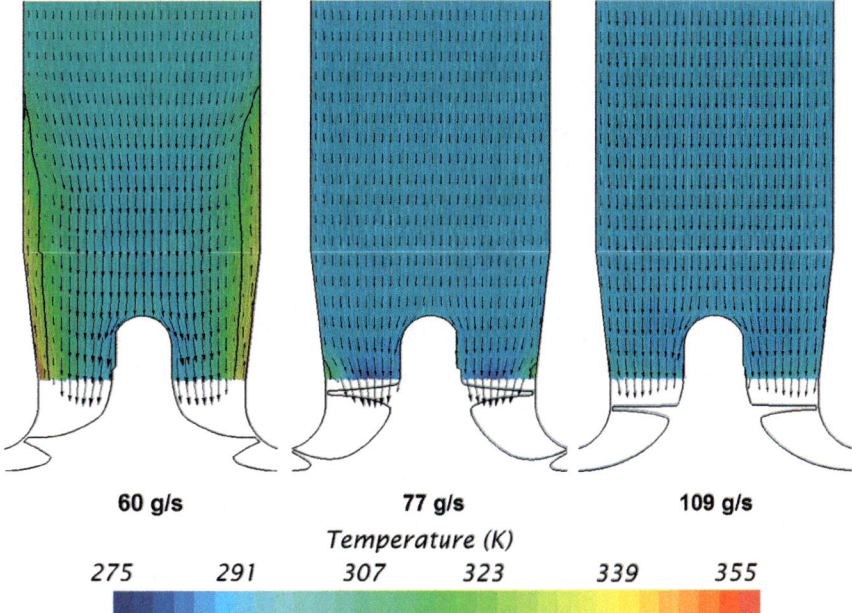

Fig. 5.2 LIC of velocity blended with temperature contours with velocity vectors at plane section for different operating conditions

radius from the wall. As the probes are placed further upstream, the temperature rise is diminished. Sensors located about two diameters upstream the inducer plane detect a temperature increase only in operating points close to surge, so they would be the best candidates to act as indicators of surge onset. Torregrosa et al. [8] noted that the upstream distance reached by the temperature increase (and thus recirculations) grows as the working point is shifted towards lower mass flow rates, being about two inlet diameters in the last stable point before surge onset. Dehner et al. [9] captured time-resolved temperature signals during mild and deep surge using fast-response thermocouples placed at the inlet duct axis. The sensor located at the inducer plane detects large temperature oscillations, occurring at the dominant frequency of the pressure pulsations. The temperature probe placed two diameters upstream the inducer plane also detects a temperature rise with this frequency, but its magnitude is greatly reduced.

Figure 5.3 depicts the flow at the inducer plane for the different cases. In-plane velocity vectors are displayed over contours of axial velocity. At this plane, proximity of main blades leading edge causes a reduction of axial velocity. In-plane vectors are displayed keeping the same scale at different operating conditions so as to allow a safe comparison. At 109 g/s, flow enters axially into the inducer, which can be concluded by noticing the small size of the in-plane vectors. Conversely, the other two operating points present secondary flows at the inducer plane. Backflow region presents high angular momentum in the clockwise direction, which, as seen in Chap. 3, is imparted

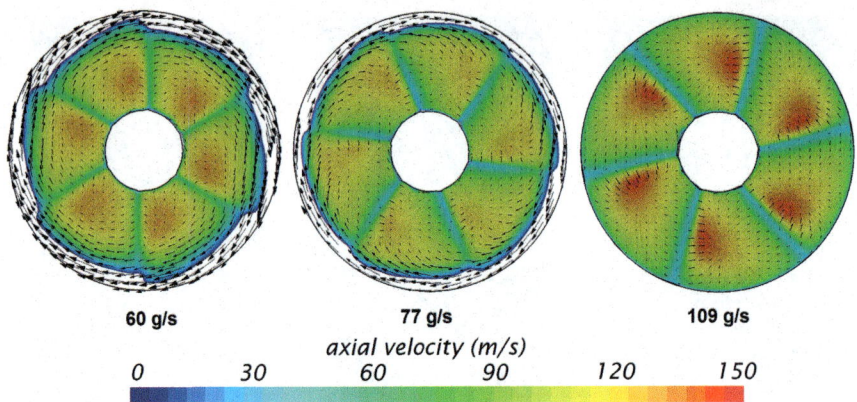

Fig. 5.3 LIC of velocity blended with axial velocity contours with velocity vectors at inducer plane for different operating conditions

by the impeller. The shear stress created by the recirculating flow creates positive pre-rotation in the incoming flow, thus improving the incidence angle. This effect extends to midspan location at 60 g/s, being more limited in 77 g/s.

Besides, the recirculation shown in Fig. 5.2 increases axial velocity of the incoming flow because it reduces the available cross-section, i.e., the fraction of the inducer plane with positive values for axial velocity in Fig. 5.3 diminishes as mass flow is reduced. At 77 g/s, backflow turning point is too close to the inducer plane, and secondary flow from shroud to hub is detected at this plane. However, recirculation flow in 60 g/s case reaches further upstream (see left side of Fig. 5.2) and the axial velocity is higher than at 77 g/s even though mass flow rate is lower. In this way, 60 g/s case presents warmer colors than 77 g/s in Fig. 5.3.

In the light of these results, 60 g/s operating point should present better incidence angle than 77 g/s, because both axial and positive tangential velocity are higher. This increase of stability produced by the backflow at the inlet duct periphery would come at the expense of compressor isentropic efficiency.

5.5   Impeller Passages

Figure 5.4 depicts LIC of relative velocity colored by meridional velocity with relative velocity vectors on top for the different operating conditions at midspan. Section A.3.1 describes the approach followed to obtain the meridional velocity scalar field. All variables have been time-averaged over several impeller revolutions.

At 109 g/s, the flow arrives with a positive incidence to the leading edge of the main blades. In the upper passage (please refer to A.2 for clarification on the nomenclature), the jet is directed to the splitter blade PS because of Coriolis force. The velocity deficit at the main blade SS cannot be referred as a boundary layer (BL)

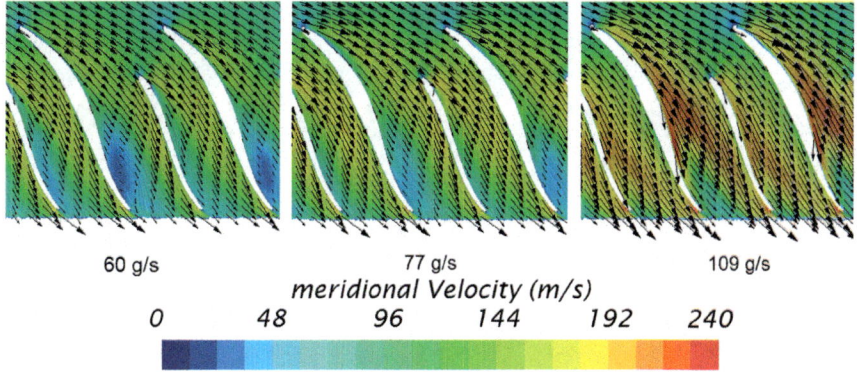

Fig. 5.4 LIC of relative velocity blended with meridional velocity contours with relative velocity vectors at impeller midspan for different operating conditions

separation, because a stream of flow with high velocity remains in the SS until the trailing edge. In the lower passage, the jet starts to migrate to the main blade PS just upstream the splitter blade trailing edge. Therefore, the lower passage delivers more mass flow than the upper channel, as noted by Després et al. [2].

At 77 g/s, the angle of incidence is more positive because the axial component is reduced due to the decrease in mass flow rate. In this way, the BL at the inducer SS is thicker than for 109 g/s. In the upper passage, the BL detaches at main blade midchord position, creating a low momentum region close to the blade SS that reduces the effective cross-section area. In the lower passage, BL separation is also present, being less severe than in the upper one.

At 60 g/s, the incidence angle is better than at 77 g/s, as expected by the results shown in Sect. 5.4. However, the boundary layer separation occurs at same chordwise location as in 77 g/s. Moreover, stall at main blade SS is more intense than at 77 g/s, i.e., meridional velocity is lower in this region.

Figure 5.5 shows LIC of relative velocity blended with pressure contours at 50% span. All 3 operating conditions are alike: pressure rise is built up in the exducer, where the passage becomes radial. The stagnation point of the blade leading edge presents a local maximum in pressure, followed by a sudden decrease in pressure in the first section of the blade SS. This low-pressure region is more patent in 109 g/s, due to the higher velocity of the flow.

Figure 5.6 shows LIC of relative velocity combined with temperature contours at midspan. At 109 g/s, the zone of low pressure next to the main blade stagnation point is also characterized by low temperature. In fact, temperatures of about 250 K are achieved in this operating point. Heat transfer between the fluid flow and the impeller, which is not considered, should prevent the temperature from dropping to these unrealistic low values. Temperature field at impeller midspan almost matches pressure field depicted in Fig. 5.5 save for one aspect: flow in 60 g/s case is already hot at the inducer. This is due to the effect of the recirculating flow, as explained in Sect. 5.4.

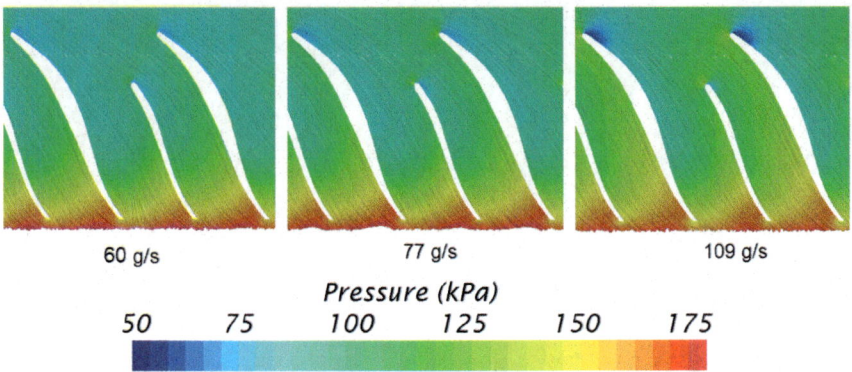

Fig. 5.5 LIC of relative velocity blended with pressure contours at impeller midspan for different operating conditions

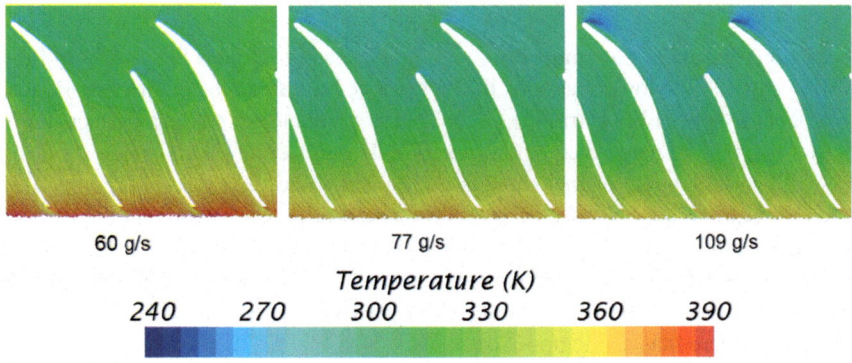

Fig. 5.6 LIC of relative velocity blended with temperature contours at impeller midspan for different operating conditions

Figure 5.7 presents the flow behavior at surfaces located at midpitch distance between blade PS and SS for different operating conditions. On top of Fig. 5.7 midpitch surfaces corresponding to upper passage are shown, whereas the sections corresponding to the lower passages are shown at bottom. LIC of relative velocity is blended with contours of meridional velocity, with the addition of relative velocity vectors. The change from incoming flow in the compressor core (positive meridional velocity) to recirculating flow near the shroud (negative meridional velocity) is indicated by a solid black line.

At 109 g/s, recirculation in the upper passage is limited to the blade projection. The lower passage does not suffer from flow reversal; only boundary layer growth in the streamwise direction is noticed. In this way, the flow is much more evenly distributed in the lower passage than in the upper passage, which is dominated by a jet located at midspan position. The case with mass flow rate of 77 g/s presents backflow that goes upstream the inducer plane but does not reach the impeller's

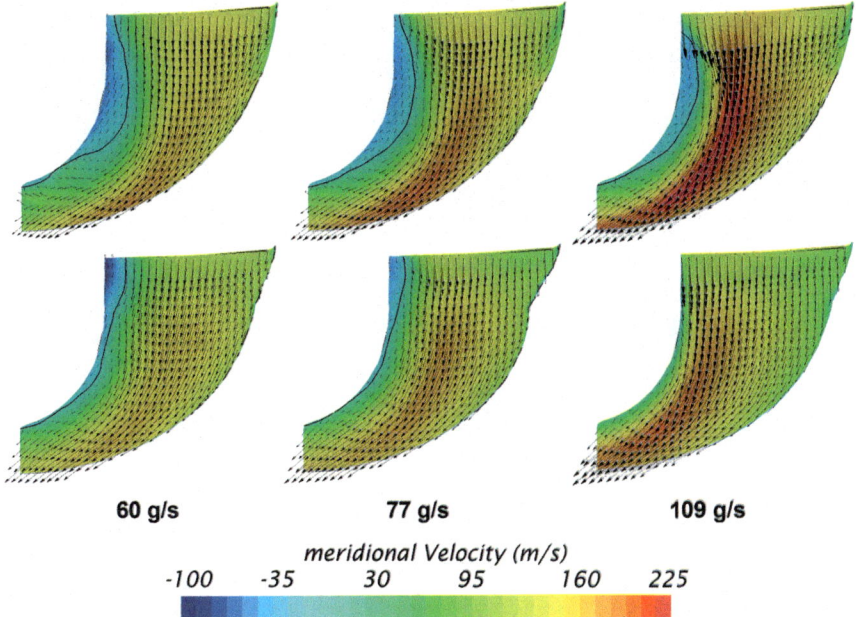

Fig. 5.7 LIC of relative velocity blended with meridional velocity contours with relative velocity vectors at midpitch surfaces corresponding to upper passage (top) and lower passage (bottom), for different operating conditions

eye (see Fig. 5.2). The lower passage at these operating conditions does suffer from flow reversal, although backflow region is not as thick as in the upper passage. The recirculating flow and the flow that reincorporates to the main stream affect to a greater fraction of compressor span, thus displacing the main jet to the hub. Finally, at 60 g/s the flow pattern is quite similar to the one described for 77 g/s, although the characteristics are more prominent due to the reduction in mass flow rate.

Figure 5.8 presents LIC of relative velocity combined with pressure contours. For all operating conditions, pressure field is almost uniform in a streamwise coordinate at the radial part of the passages. However, the axial part of the passages is characterized by a low-pressure stream located between the locus of zero meridional velocity (solid black line) and the main jet observed in Fig. 5.7. This phenomenon is more noticeable at upper passages.

Figure 5.9 displays temperature contours over LIC of relative velocity. Apart from the increase of temperature associated with the pressure rise in the exducer depicted in Fig. 5.8, the effect of the recirculating flow in the internal heat transfer dominates the temperature field at these midpitch surfaces. Backflow is hotter than the freestream incoming flow because it has been previously compressed. The flow that is adjacent to backflow also presents high temperature because it comes from the recirculating flow that reintegrates back into the main stream. Moreover, secondary flows and high turbulence increase the diffusive heat transfer in the spanwise direction due to the

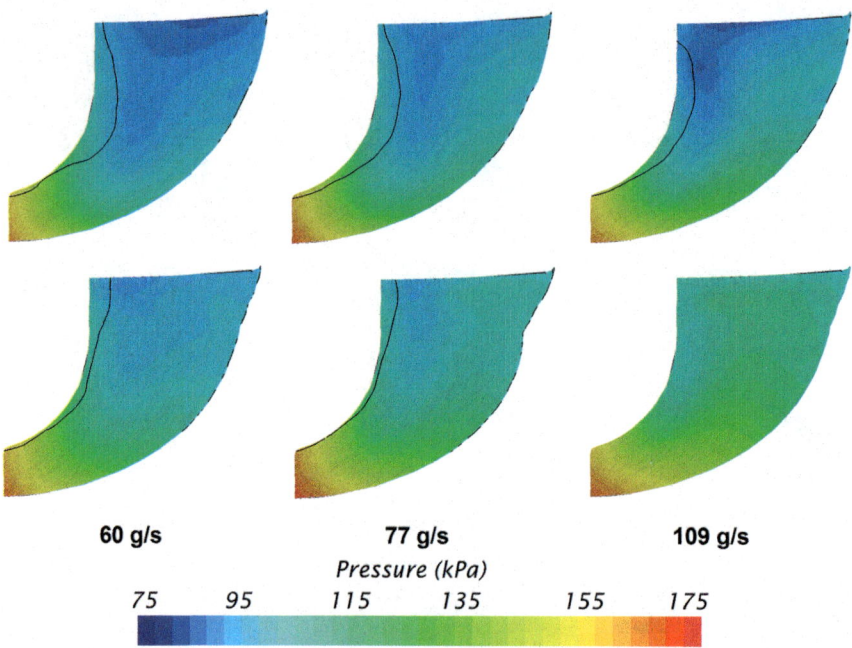

Fig. 5.8 LIC of relative velocity blended with pressure contours at midpitch surfaces corresponding to upper passage (top) and lower passage (bottom), for different operating conditions

increase of the mixing related to integral and smaller length scales. Besides, a local stream of low temperature can be observed in the upper passage at 109 g/s, which corresponds to the high-velocity jet described in Fig. 5.7.

In order to complete the overall portrait of the flow at the impeller passages, a set of 4 isomeridional cross-section will be used in the next pictures. Figure 5.10 presents flow at these postprocessing surfaces for the different operating conditions. Relative velocity LIC is colored with meridional velocity and combined with relative velocity vectors. A solid black line indicates zero meridional velocity.

The onset of recirculating flow when the compressor works at 109 g/s is clearly located at the main blade leading edge SS. The positive incidence angle is responsible for the inception of the backflow, which spreads into the circumferential direction until reaching the splitter blade PS in the 3rd cross-section. Flow pattern at 77 and 60 g/s is alike except at the 1st plane. At 77 g/s the incoming flow close to the recirculating region moves from shroud to hub because the turning point of the backflow is at a short axial distance (see Fig. 5.2), whereas the case with 60 g/s does not suffer from this phenomenon since the recirculating flow reintegrates much further upstream. Finally, flow migration from shroud to hub is more intense in the lower passages, as can be noticed in the 3rd and 4th cross-sections of Fig. 5.10 for all operating conditions.

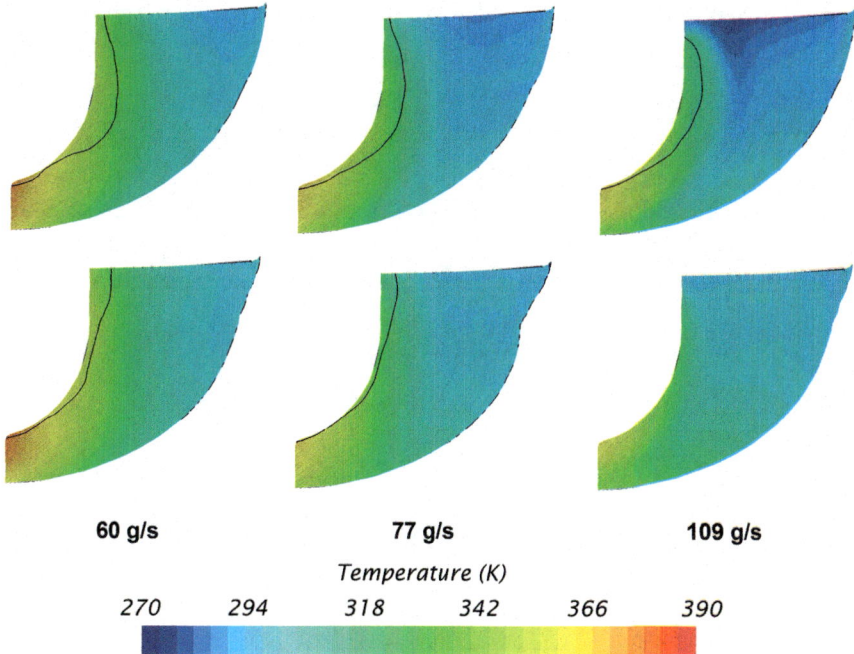

Fig. 5.9 LIC of relative velocity blended with temperature contours at midpitch surfaces corresponding to upper passage (top) and lower passage (bottom), for different operating conditions

Figure 5.11 presents pressure contours combined with relative velocity LIC at the meridional surfaces. As observed in previous pictures, pressure rise is mainly attained in the last cross-sections for all operating conditions. Blade loading is higher in the main blades than in the splitter ones, which is derived from the observation of pressure jump between blade PS and SS. This characteristic results in a BPF tone corresponding to the main blades instead of the whole set of blades, as will be seen in Chap. 6.

A pressure increase in the spanwise direction from shroud to hub can also be observed in Fig. 5.11, particularly at 109 g/s. This feature is confirmed by looking again at pressure meriodional distribution depicted in Fig. 5.8. Flow leaving the main blade leading edge produces local pressure minimum where the angle of attack is more positive, which again is more prominent for 109 g/s mass flow rate.

Figure 5.12 shows LIC of relative velocity blended with temperature contours. It is confirmed that temperature field does not depend on circumferential coordinate for other isospan positions apart from the one studied in Fig. 5.6. Only the low temperature stream described in Fig. 5.9 for the case with 109 g/s presents a pitchwise variation in the 1st and 2nd cross-sections of Fig. 5.12.

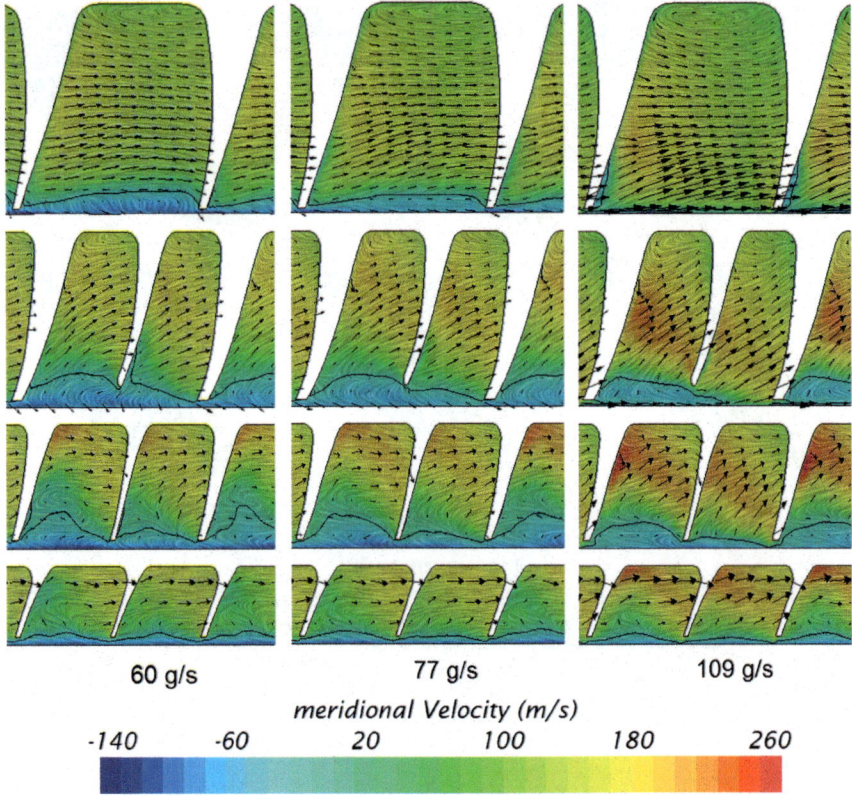

Fig. 5.10 LIC of relative velocity blended with meridional velocity contours with relative velocity vectors at a set of isomeridional surfaces for different operating conditions

5.6 Diffuser, Volute and Outlet Duct

Mean values obtained at the diffuser should be interpreted with care because this region is included into the rotating mesh. Therefore, the time-averaged fields depicted at the diffuser do not reflect the unaxisymmetry of the volute but the periodic structures formed by the jet-wake pattern in the impeller passages.

Figure 5.13 represents cross-sections of the diffuser at isoazimuthal angle. The location of the plane corresponds to midptich distance at the upper passage, taking into account that the relative position to the impeller is the one that matters instead of the location regarding the volute, because this region is included into the rotating mesh. In-plane velocity vectors are displayed over LIC of absolute velocity combined with contours of radial velocity for the three studied operating conditions. Zero radial velocity is indicated by a black solid line. The upper wall of the diffuser in Fig. 5.13 corresponds to the continuation of the impeller shroud, whereas the bottom wall is the diffuser hub.

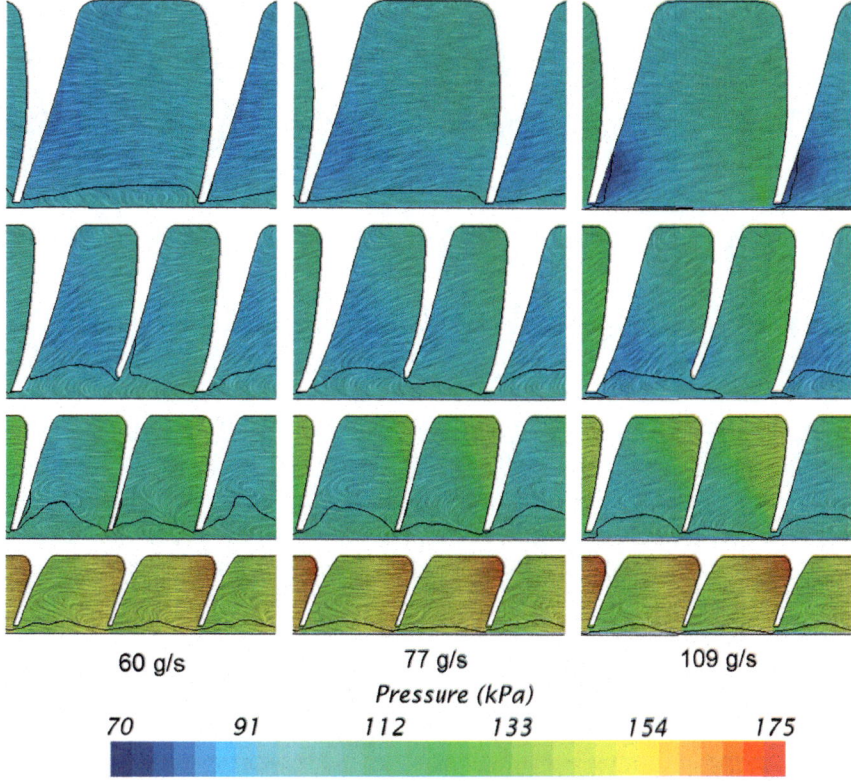

Fig. 5.11 LIC of relative velocity blended with pressure contours at a set of isomeridional surfaces for different operating conditions

At 109 g/s, high radial velocity exists just downstream the impeller. Conservation of mass flow rate dictates that radial velocity should be decreased as diffuser radius is increased, since cross-sectional area is increased as well because diffuser width is constant. Boundary layer growth can also be observed at this operating point, but no recirculating flow appears. Radial velocity is lower for the case with 77 g/s mass flow rate, and backflows at the diffuser hub (the bottom wall in Fig. 5.13) exist. Near surge, the flow in the diffuser is similar to the case at maximum compression ratio (Fig. 5.13 top and middle, respectively). The only difference is that a thick zone of reverse flow appears at the bottom wall of the diffuser for 60 g/s mass flow rate. As shown in Fig. 5.3, recirculating flow allows the compressor to be stable near surge, by reducing cross-sectional area and depicting a similar flow field as in the case of 77 g/s but with a reduced incoming flow area. Després et al. [2] also found a recirculation region at diffuser hub.

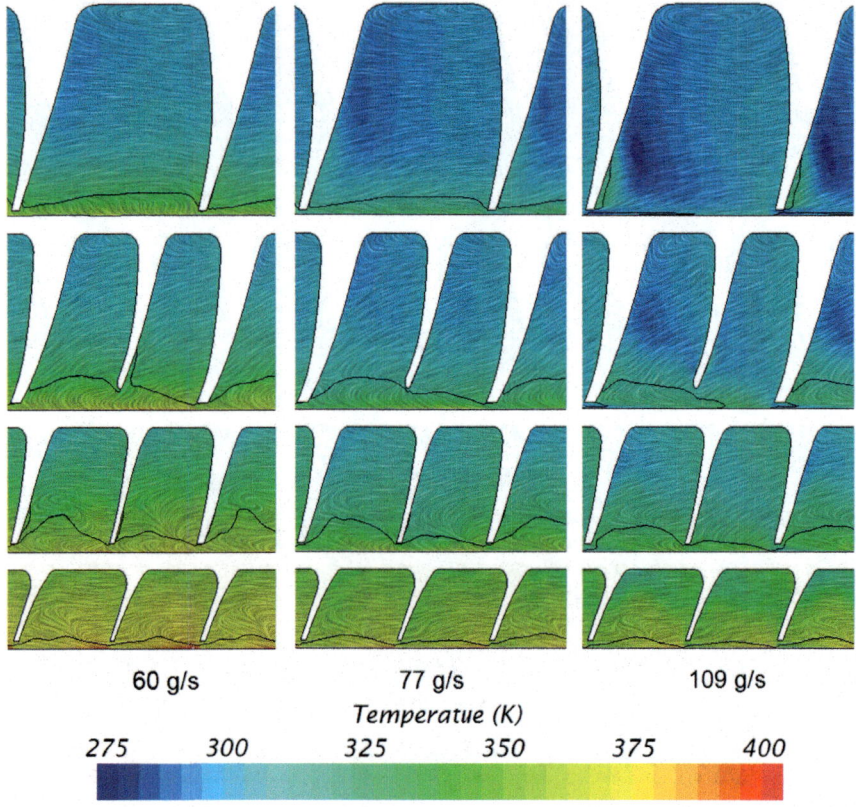

Fig. 5.12 LIC of relative velocity blended with temperature contours at a set of isomeridional surfaces for different operating conditions

Figure 5.14 shows the diffuser plane at 50% span. LIC of absolute velocity field is blended with contours of velocity angle, defined as

$$\alpha = \arctan\frac{v_{rad}}{v_{tang}}, \tag{5.1}$$

i.e., lower velocity angles correspond to flow more oriented towards tangential direction. Several streamlines are superimposed as an estimation of fluid motion.

At 109 g/s, jet-wake structure can be clearly seen in terms of high-low velocity angle contours. In the relative reference frame, jet-wake pattern is not mixed at the diffuser, obtaining an average angle of 40° at diffuser outlet at BEP. 77 g/s mass flow rate case shows a better mixing of the set of jets and wakes, because radial velocity is smaller than at 109 g/s. Velocity angle is thus decreased, which can be confirmed by the cooler contours or by noticing that the absolute velocity streamlines are more

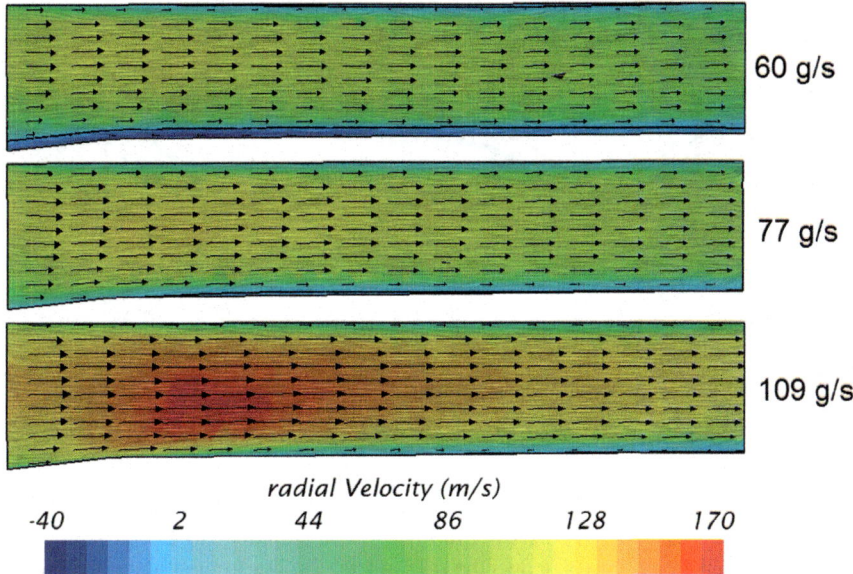

Fig. 5.13 LIC of velocity blended with contours of radial velocity with velocity vectors at a diffuser isoazimuthal surface for different operating conditions

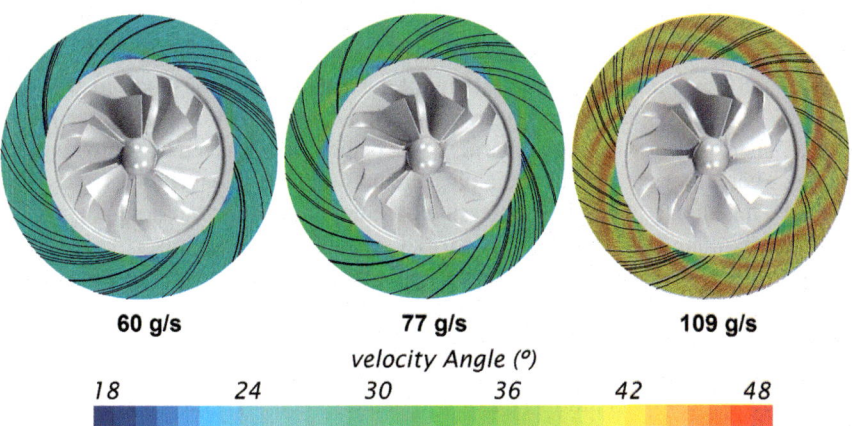

Fig. 5.14 LIC of velocity blended with contours of velocity angle with streamlines at diffuser midspan for different operating conditions

oriented towards tangential direction. Near surge, the flow is even more tangential and the jet-wake structure is only visible just upstream the blades trailing edges.

In order to obtain time-averaged flow at the diffuser outlet which reflects the circumferential profile of the flow arriving to the volute, a circumference is defined at 50% span with a radius slightly higher than the one corresponding to the

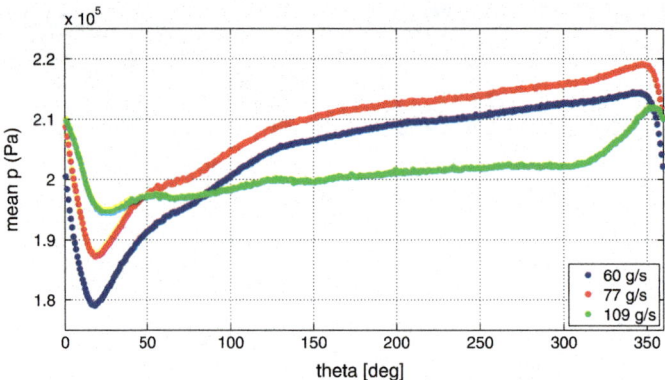

Fig. 5.15 Azimuthal distribution of pressure at volute inlet (50% span) for different operating conditions

diffuser-volute interface. In this way, the cells corresponding to this postprocessing circumference are not related to the diffuser region but to the volute one, and a meaningful time-averaging can be performed.

Figure 5.15 presents the azimuthal pressure profile at the diffuser outlet for the studied operating conditions. Volute tongue is located at 0° and volute cross-section increases with circumferential coordinate. The case with highest mass flow rate presents a constant adverse pressure gradient in volute streamwise direction followed by a sudden rise and fall in pressure. Point of maximum pressure is located at volute tongue. 77 and 60 g/s cases present similar profile. A pressure drop exists after volute tongue. Then, pressure increases sharply at the beginning less steep just before volute tongue. There only exists an offset in pressure between the cases with less mass flow rate, which is due to the different pressure at outlet boundary condition.

Jyothishkumar et al. [1] also found that pressure rise at diffuser periphery is gradual at BEP, whereas a low-pressure region after the volute tongue dominates the circumferential profile near surge. Similar conclusions are obtained in the works of Guo et al. [3] and Chen et al. [4].

Figure 5.16 depicts radial velocity azimuthal distribution at volute inlet. For all the studied operating conditions, maximum radial velocity is acquired at 50°. Radial velocity is quite uniform at BEP whereas circumferential profile for cases with less mass flow rate is dominated by a region of high radial velocity after volute tongue.

Chen et al. [4] showed similar profiles for low mass flow rate operating conditions. However, point with design flow also displayed a marked zone with high radial velocity. This working point was calculated with a steady simulation. Circumferential profiles obtained in this way reflect the jet-wake structure imposed by the constant position of the blades, lacking of the azimuthal smoothness that confers the time-averaged motion of the impeller. Circumferential distributions provided by steady-state simulation at BEP are not consistent with the rest of the transient simulations showed by Chen et al. [4] and differ from the one shown in Fig. 5.16. Therefore, steady

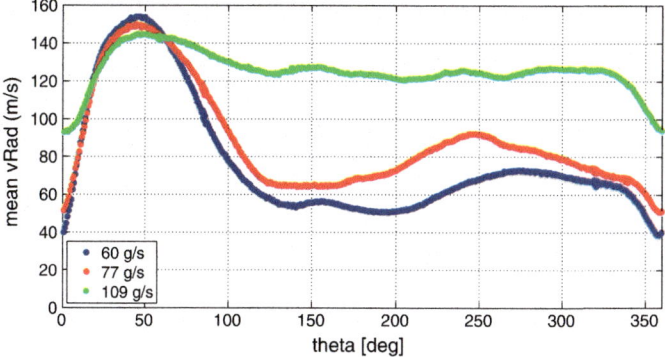

Fig. 5.16 Azimuthal distribution of radial velocity at volute inlet (50% span) for different operating conditions

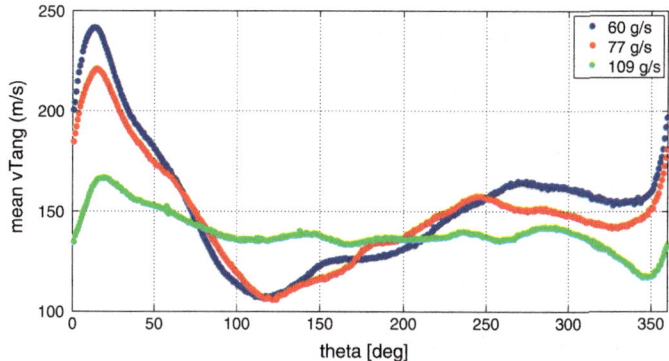

Fig. 5.17 Azimuthal distribution of tangential velocity at volute inlet (50% span) for different operating conditions

simulations may not be suitable for providing azimuthal profiles at the diffuser, even at design mass flow rate. Bousquet et al. [10] compared steady and transient simulations, showing a difference in behavior at the vaneless space of a vaned diffuser, with increasing discrepancy as mass flow is reduced. Zheng et al. [11] also claimed that steady state simulations using frozen rotor approach should be interpreted with care when predicting compressor flow field circumferential distributions with asymmetric volutes.

Figure 5.17 presents circumferential profile of tangential velocity. A similar description than the one provided for Fig. 5.16 holds: maximum tangential velocity is achieved after volute tongue and azimuthal distribution at 109 g/s is much smoother than the ones obtained at low mass flow rate conditions. Nevertheless, case at BEP displayed a higher radial velocity than the other two cases at almost all circumferential positions (see Fig. 5.16) whereas mean tangential velocity at diffuser outlet is higher as mass flow rate is reduced.

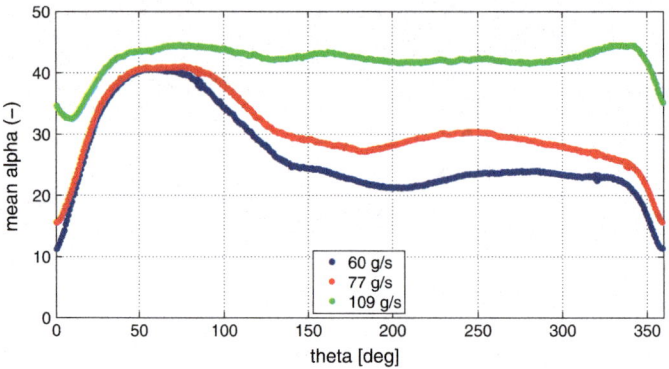

Fig. 5.18 Azimuthal distribution of velocity angle at volute inlet (50% span) for different operating conditions

Figure 5.18 displays the azimuthal distribution of flow velocity angle defined by Eq. 5.1. Evolution of mean velocity angle at diffuser outlet has been already discussed in Fig. 5.14: flow is more oriented towards tangential direction (velocity angle is reduced) with decreasing mass flow rate. Figure 5.18 shows that case with highest mass flow delivers flow to the volute with an almost constant velocity angle. Conversely, working points closer to surge present a good velocity angle only at the region corresponding to high radial velocity (see Fig. 5.16). Outside this zone, flow is more aligned with circumferential direction, particularly before volute tongue.

After having analyzed the flow conditions at volute inlet, the evolution of time-averaged flow field at scroll and outlet duct will be studied. To do so, a set of 5 cross-section surfaces will be displayed in streamwise sequence for the different operating conditions.

Figure 5.19 depicts in-plane velocity vectors superimposed to LIC of velocity field blended with in-plane velocity contours. Flow at first two surfaces is alike for all operating conditions, because incoming radial velocity is similar at this zone (see Fig. 5.16). As radial velocity is reduced for cases with less mass flow rate, the characteristic swirling motion of the cross-section presents less intensity than at BEP, which is patent in 3rd to 5th surfaces of Fig. 5.19. Therefore, the concentrical vortical pattern is distorted as compressor working point is shifted towards surge, increasing the possibility of scroll vortex breakdown.

Chen et al. [4] showed that several small vortices appear at the center of volute cross-section when vortex breakdown happens for operating points close to surge. Measurements performed by Hagelstein et al. [12] displayed a counter-rotating vortex in a volute operating near surge. Swirl motion was only observed for higher mass flow rates. Hagelstein et al. also performed simulations of the volute alone with a steady, inviscid solver with a term to account for wall friction effects and internal shear. However, their code failed to predict the onset of vortex breakdown for near-surge operating conditions. Pitkanen et al. [13] proved that volutes with sharp corners

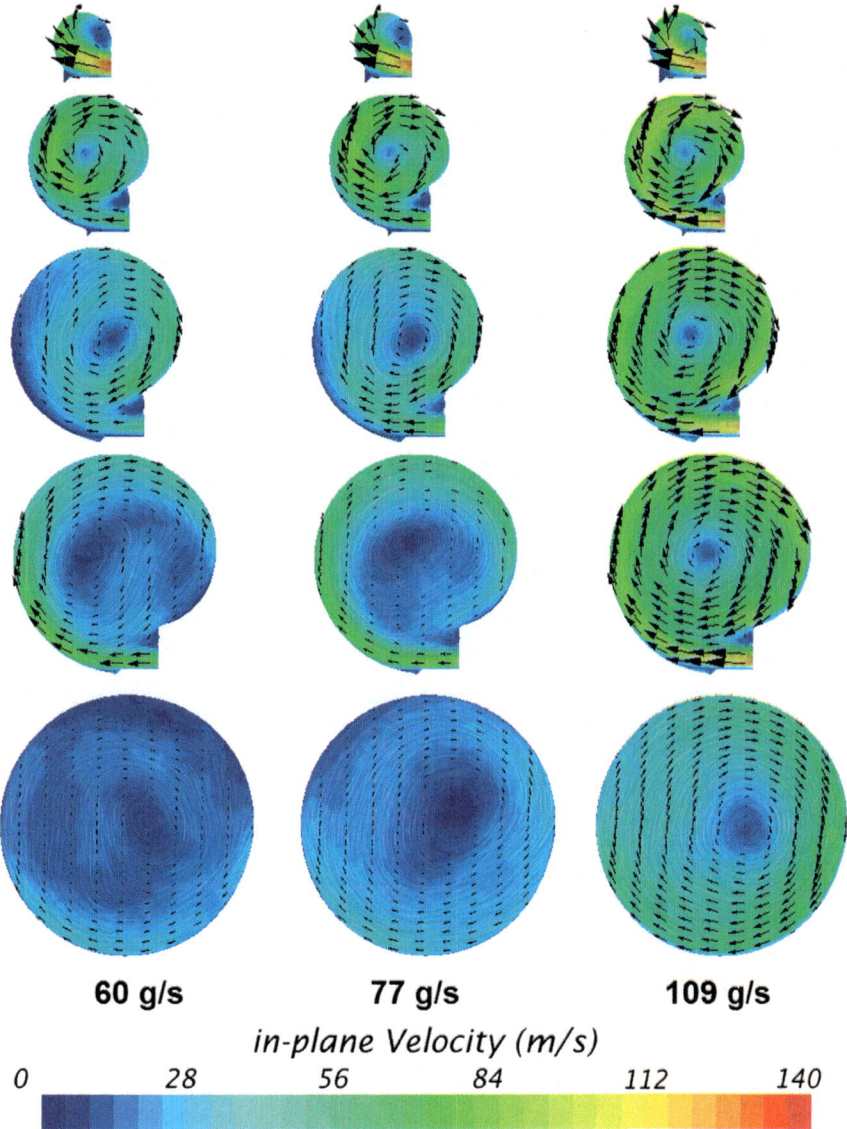

Fig. 5.19 LIC of velocity blended with in-plane velocity contours with velocity vectors at volute cross-section surfaces for different operating conditions

are more prone to develop secondary vortices than volutes based on a circular cross-sectional shape, such as the one studied in the present work.

Figure 5.20 presents circumferential (normal) velocity contours combined with LIC of velocity field. Contrary to what might be expected, circumferential velocity at first cross-section surface is higher as working point is closer to surge, although this

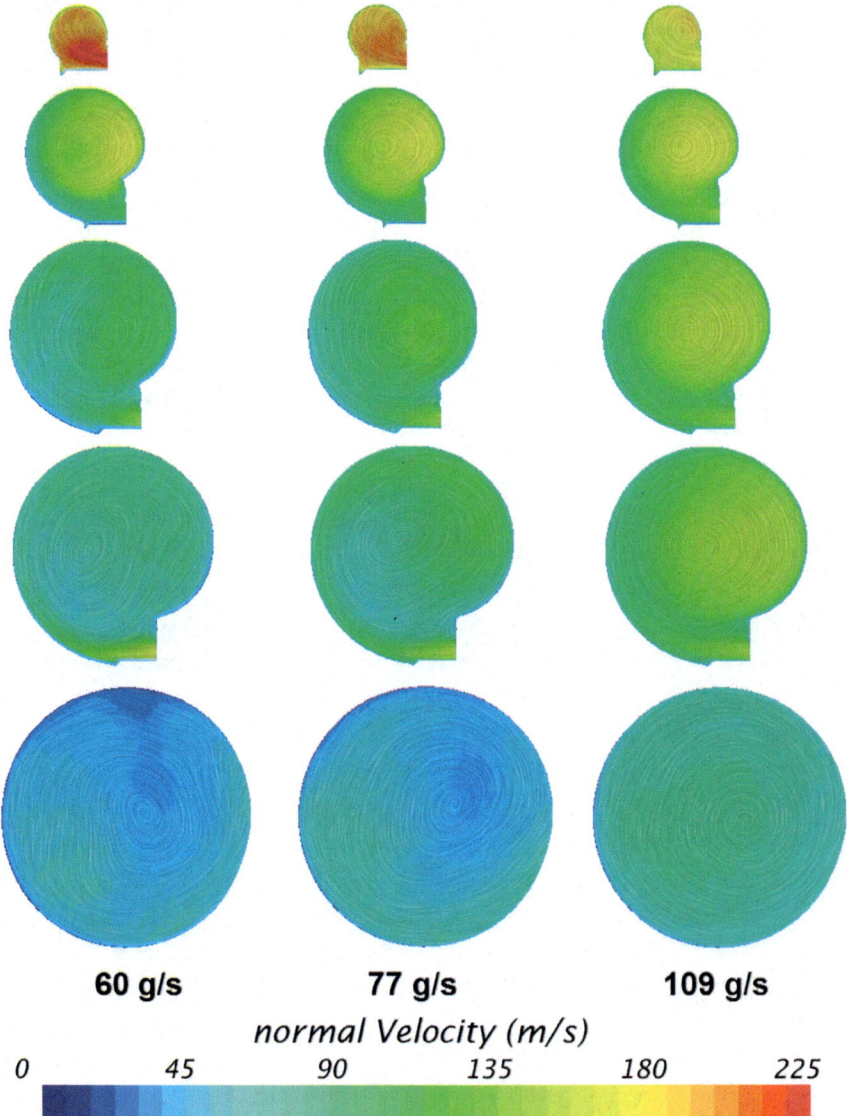

Fig. 5.20 LIC of velocity blended with normal velocity contours at volute cross-section surfaces for different operating conditions

information could also be extracted from Fig. 5.17. Azimuthal velocity is similar for all operating conditions at 2nd surface, whereas, from 3rd section onwards, normal velocity is higher with increasing mass flow rate.

In the scroll, both cross-sectional area and mass flow rate increase in the streamwise direction. In any case, area growth is greater than mass flow increase, and the volute acts as as diffuser thus reducing normal velocity in the streamwise direction for all operating conditions, as depicted in Fig. 5.20. Observation of Fig. 5.17 may have led to a spurious conclusion in terms of normal velocity distribution in the scroll. Figure 5.17 is obtained using a virtual circumference placed at the volute inlet. This location is not representative of the mean normal velocity at the volute, because volute inlet tangential velocity is higher than the surface-averaged tangential velocity at 3rd and 4th cross-sections of Fig. 5.20, particularly for cases with 60 and 77 g/s mass flow rate.

5.7 Conclusions

In this chapter, the evolution of mean flow through the turbocharger compressor has been described. Three operating conditions at 160 krpm compressor speed are studied. Starting from best efficiency point (109 g/s) at this isospeed, mass flow rate is reduced until maximum compression ratio is obtained (77 g/s), decreasing even more the mass flow to achieve near-surge conditions (60 g/s). Pressure ratio, specific work and isentropic efficiency are predicted with relative differences compared to experiments below 3%.

At 109 g/s, flow is more *regular* than at the other two working points, i.e., secondary flows are less intense, the flow is more axisymmetric at diffuser outlet, etc. Interestingly, the case with 109 g/s presents recirculating flow, although it is confined to the impeller upper passage.

Flow at the other two working points (77 and 60 g/s) is alike. This is because case with less mass flow presents thicker backflow zone near the walls, thus presenting a flow similar to the one delivered by 77 g/s case at compressor core. Particularly, inlet recirculation reaches the impeller's eye in 77 g/s, but extends 1 diameter upstream the impeller's eye for 60 g/s.

Finally, volute is dominated by swirling flow at 109 g/s, which degrades at 77 g/s and is about to break down for 60 g/s.

Once the main features of the time-averaged flow at different operating conditions are known, the analysis of aeroacoustic phenomena at same working points can be addressed. Chapter 6 is devoted to such a task.

References

1. V. Jyothishkumar, M. Mihaescu, B. Semlitsch, L. Fuchs, Numerical flow analysis in centrifugal compressor near surge condition, in *Fluid Dynamics and Co-located Conferences* (American Institute of Aeronautics and Astronautics, 2013), p. 13. https://doi.org/10.2514/6.2013-2730
2. G. Després, G.N. Boum, F. Leboeuf, D. Chalet, P. Chesse, A. Lefebvre, Simulation of near surge instabilities onset in a turbocharger compressor. Proc. Inst. Mech. Eng Part A J. Power Energy **227**(6), 665–673 (2013). https://doi.org/10.1177/0957650913495537

3. Q. Guo, H. Chen, X.C. Zhu, Z.H. Du, Y. Zhao, Numerical simulations of stall inside a centrifugal compressor. Proc. Inst. Mech. Eng. Part A J. Power Energy **221**(5), 683–693 (2007). https://doi.org/10.1243/09576509JPE417

4. H. Chen, S. Guo, X.C. Zhu, Z.H. Du, S. Zhao, Numerical simulations of onset of volute stall inside a centrifugal compressor. in *Proceedings of ASME Turbo Expo 2008: Power for Land, Sea and Air* (ASME, 2008)

5. H.K. Versteeg, W. Malalasekera, *An Introduction to Computational Fluid Dynamics: The Finite Volume Method*, 2nd edn. (Pearson Education Limited, Harlow, 2007)

6. B. Cabral, L.C. Leedom, Imaging vector fields using line integral convolution, in *Proceedings of the 20th Annual Conference on Computer Graphics and Interactive Techniques* (ACM, 1993), pp. 263–270. https://doi.org/10.1145/166117.166151

7. J. Andersen, F. Lindström, F. Westin, Surge definitions for radial compressors in automotive turbochargers. SAE Int. J. Engines **1**(1), 218–231 (2008). https://doi.org/10.4271/2008-01-0296

8. A. Torregrosa, A. Broatch, X. Margot, J. García-Tíscar, Y. Narvekar, R. Cheung, Local flow measurements in a turbocharger compressor inlet. Exp. Therm. Fluid Sci. **88**, 542–553 (2017). https://doi.org/10.1016/j.expthermflusci.2017.07.007. ISSN: 0894-1777

9. R. Dehner, N. Figurella, A. Selamet, P. Keller, M. Becker, K. Tallio, K. Miazgowicz, R. Wade, Instabilities at the low-flow range of a turbocharger compressor. SAE Int. J. Engines **6**(2), 1356–1367 (2013). https://doi.org/10.4271/2013-01-1886

10. Y. Bousquet, X. Carbonneau, I. Trébinjac, Assessment of steady and unsteady model predictions for a subsonic centrifugal compressor stage, in *ASME Turbo Expo 2012*

11. X.Q. Zheng, J. Huenteler, M.Y. Yang, Y.J. Zhang, T. Bamba, Influence of the volute on the flow in a centrifugal compressor of a high-pressure ratio turbocharger. Proc. Inst. Mech. Eng. Part A J. Power Energy **224**(8), 1157–1169 (2010). https://doi.org/10.1243/09576509JPE968, eprint: http://pia.sagepub.com/content/224/8/1157.full.pdf+html

12. D. Hagelstein, K. Hillewaert, R.A. Van den Braembussche, A. Engeda, R. Keiper, M. Rautenberg, Experimental and numerical investigation of the flow in a centrifugal compressor volute. J. Turbomach. **122**(1), 22–31 (1999). https://doi.org/10.1115/1.555423

13. H. Pitkanen, H. Esa, A. Reunanen, P. Sallinen, J. Larjola, Computational and experimental study of an industrial centrifugal compressor volute. J. Therm. Sci. **9**(1), 77–84 (2000). https://doi.org/10.1007/s11630-000-0047-5

Chapter 6
Compressor Aerocoustics at Near-Stall Conditions

6.1 Introduction

As stated in Chap. 1, the aim of this book is to contribute to the understanding of turbocharger compressor flow-induced acoustics, particularly regarding whoosh noise phenomenon. To do so, a methodology to compare numerical and experimental spectra has been developed in Chap. 2, which has been used as a tool to evaluate the sensitivity of the noise predicted by the model to parameters such as tip clearance, grid spacing, time-step size, etc. In Chap. 5, using the validated configuration, the main features of the time-averaged flow at three operating conditions have been described.

Finally, this chapter deals with turbocharger compressor flow-induced acoustics at three operating conditions in the surge side of 160 krpm. A literature review is performed in Sect. 6.2. Comparison of URANS and DES acoustic signature against experiments is performed in Sect. 6.3. Then, Sect. 6.4 is devoted to the investigation of flow field to identify the phenomena responsible for the acoustic signature features. Finally, chapter conclusions are included in Sect. 6.5.

6.2 Literature Review

Bousquet et al. [1] conducted a numerical study of a centrifugal compressor. Three operating points were studied from peak efficiency to near stall conditions. A virtual probe linked to the relative relative frame located at 90% span at the impeller inlet was used for spectral analysis. Signal is recorded during 6 impeller rotations after obtaining a stable state at near stall conditions, whereas 1 rotation is registered by the monitor at the other two working points. Vane BPF tone is the main feature for the two operating conditions with higher mass flow. Near stall, lower frequency content appears, particularly a tone at a frequency 6 times the rotation order. Flow investigation reveals that 6 vortices are formed a shed per revolution, thus being responsible

© Springer International Publishing AG 2018
R. Navarro García, *Predicting Flow-Induced Acoustics at Near-Stall Conditions in an Automotive Turbocharger Compressor*, Springer Theses, https://doi.org/10.1007/978-3-319-72248-1_6

of the observed tone. Frequency content below rotation order is not investigated by the authors, probably because it would have required a higher recording period.

Mendonça et al. [2] analyzed the aeroacoustics of an automotive radial compressor using DES. SPL spectra were obtained for points in both inlet and outlet ducts, showing a narrow band noise at a frequency about 70% of rotational speed, which corresponds to 2.5 kHz for the investigated compressor speed. A spiral mode propagating upstream from the compressor impeller was detected for the narrow band noise aforementioned. Leading-edge separation and stalled passages were found, along with a low momentum region that rotates at a slower speed than the wheel. Inducer rotating stall was thus regarded as the source of the narrow band noise. Tip leakage was considered to be the mechanism that allows the stalled passages to recover by pushing the low momentum region to the rotation-trailing passage.

Karim et al. [3] identified the bad incidence angle that low flow rates present at the leading edge of the compressor blades as the reason of whoosh noise. Large Eddy Simulations were performed in two different operating conditions with 5 different inlet configurations (using swirl vanes, short and large steps and combinations of these elements). SPL integrated over 6–12 kHz obtained at the compressor inlet was compared for the different inlet configurations. The combination of a large step with swirl vanes provided less SPL for both operating conditions. Experimental measurements with short and large step were performed in a powertrain dynamometer semi-anechoic cell. SPL integrated between 4 and 12.7 kHz using radiated noise measurements showed the superior performance of the large step over the small step.

Lee et al. [4] conducted an experimental and numerical study of the noise radiated by a turbocharger. A turbocharger test rig was mounted on an anechoic chamber, and noise was measured with a freefield microphone placed at 0.8 m from the inlet of the compressor. Noise spectra for three rotational speeds were obtained. A narrow band was found at a frequency about 3 times the rotational speed, corresponding to frequencies between 3.5 and 5.5 kHz. This was the dominant frequency in most configurations, and was attributed to the experimental setup. Besides, CFD transient calculations of the turbocharger compressor were performed. Unsteady data for one revolution was extended by repetition for acoustic analysis. Acoustic analogy was used to obtain noise spectra. The numerical approach overpredicted the measured noise spectrum and was not able to detect the narrow band component.

Fontanesi et al. [5] performed detached eddy simulations of a turbocharger compressor with a compressor by-pass valve (CBV). Two operating points at the same isospeed were investigated: one close to surge and the other at a two times higher mass flow rate. At low frequency range (below 5 kHz), simulation with higher mass flow provides a lower overall noise than the point close to surge. A narrow band noise at 2500 Hz stands out in both spectra, which is attributed to periodic flow detachment and re-attachment at the CBV junction. Experimental measurements also detected the aforementioned narrow band.

6.3 Comparison of Acoustic Signatures

URANS and DES simulations are performed with the CFD model obtained in Chap. 4, for cases with 60, 77 and 109 g/s mass flow rates at 160 krpm. For each operating condition, pressure signals are recorded after the achievement of a steady-state during no less than 120 impeller revolutions, corresponding to 50 ms. Experimental measurements, obtained according to the approach described in Sect. 2.2, are used as a means of comparison. PSD are calculated following the methodology developed in Sect. 2.4.

Figures 6.1, 6.2, 6.3, and 6.4 show spectra obtained in the aforementioned way.

Experimental inlet PSD at low frequency range (Fig. 6.1) is almost flat for BEP, showing two humps centered at 700 and 2700 Hz. Both turbulence models accurately provide the same spectrum behavior at this operating condition, although DES over-predicts the amplitude by about 5 dB. Spectrum measured for 77 g/s mass flow rate is flat, being about 7 dB greater in average than 109 g/s PSD. 77 g/s spectra is almost identical for both turbulence approaches, and mimics experimental PSD with an offset of 5 dB. Finally, 60 g/s spectrum decreases with frequency, intersecting 77 g/s PSD at 1500 Hz. For the point closest to surge (60 g/s), numerical spectra does not depend on turbulence model until 1800 Hz, where URANS overpredicts the negative slope of the PSD.

Outlet measured PSD at plane wave range (Fig. 6.2) presents a broadband elevation from 1.5 to 3 kHz near BEP (109 g/s), followed by a small hump centered at 4400 Hz. URANS provides two broadband elevation in ranges of 1.8–3 kHz and 3.2–4.7 kHz, with an offset of 7 dB. Conversely, DES PSD for 109 g/s is flat, overpredicting amplitude by 5–10 dB. Measured 77 g/s spectrum includes a broadband elevation from 0.7 to 2.5 kHz, accurately reproduced by both turbulence models alike, with an amplitude offset of less than 5 dB. Experimental PSD presents humps at 3500 and 4000 Hz, which are interpreted by numerical simulations as a broadband noise elevation from 3 to 4.8 kHz. Near surge, experimental spectrum decreases until 3500 Hz, presenting a hump at 4000 Hz and then a constant value. Conversely, URANS is unable to predict the change of PSD behavior at 3500 Hz, thus providing a decreasing spectrum. DES does reflect this change of PSD evolution, although it includes a broadband elevation from 1 to 3 kHz that is not present in the measured spectrum.

Spectra depicted at Fig. 6.2 shows two broadband elevations in almost every case: one at a frequency from 50 to 90% of rotation speed and another from 120 to 170%, which can be related to the whoosh noise phenomenon described in Sect. 6.1. These ranges agree with the features observed at the works by Mendonça et al. [2] and Fontanesi et al. [5], despite the differences between sound speeds at outlet duct (caused by different pressure ratios). This fact suggests that whoosh noise is an aerodynamic sound not related to cavity resonances. Using the terminology of Mongeau et al. [7], whoosh noise is not a Helmholtz effect but a Strouhal one. Particularly, compressor speed plays a role in the onset of whoosh noise, as indicated by Evans and Ward [8]. In fact, Després et al. [9] did not found any rotating stall at an operating point close to surge but at low compressor speed. At this isospeed, the compressor map of Després et al. did not present a positive slope region, which is

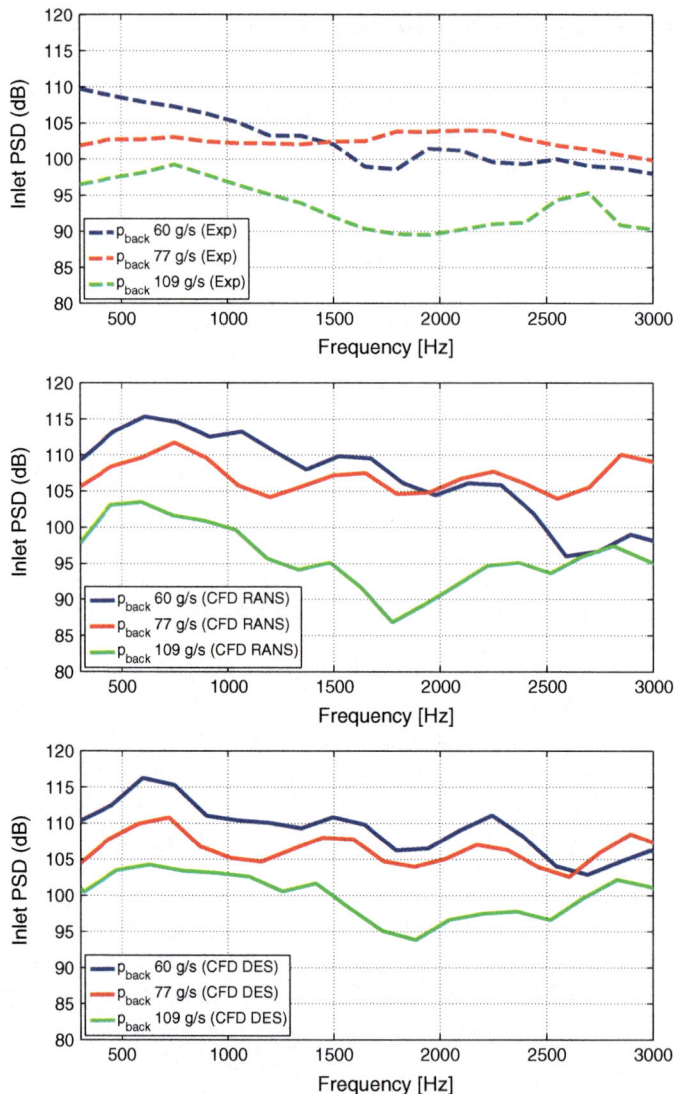

Fig. 6.1 Low frequency PSD of experimental (top), numerical URANS (middle) and DES (bottom) pressure components at inlet duct, for the different operating conditions (reprinted from [6], with permission from Elsevier)

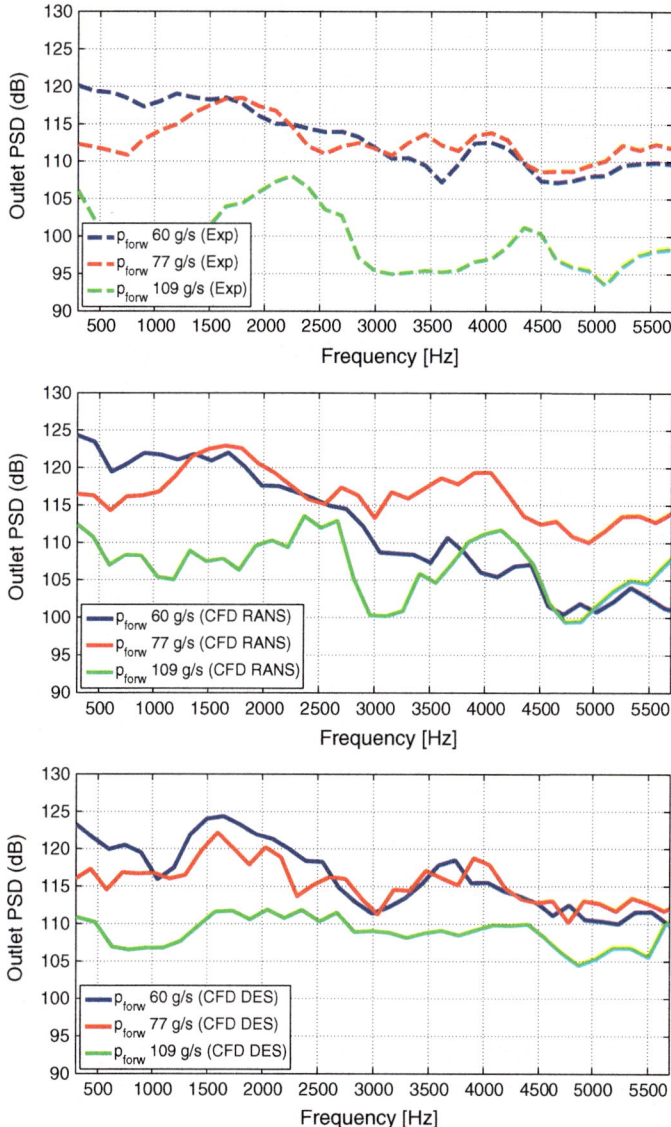

Fig. 6.2 Low frequency PSD of experimental (top), numerical URANS (middle) and DES (bottom) pressure components at outlet duct, for the different operating conditions (reprinted from [6], with permission from Elsevier)

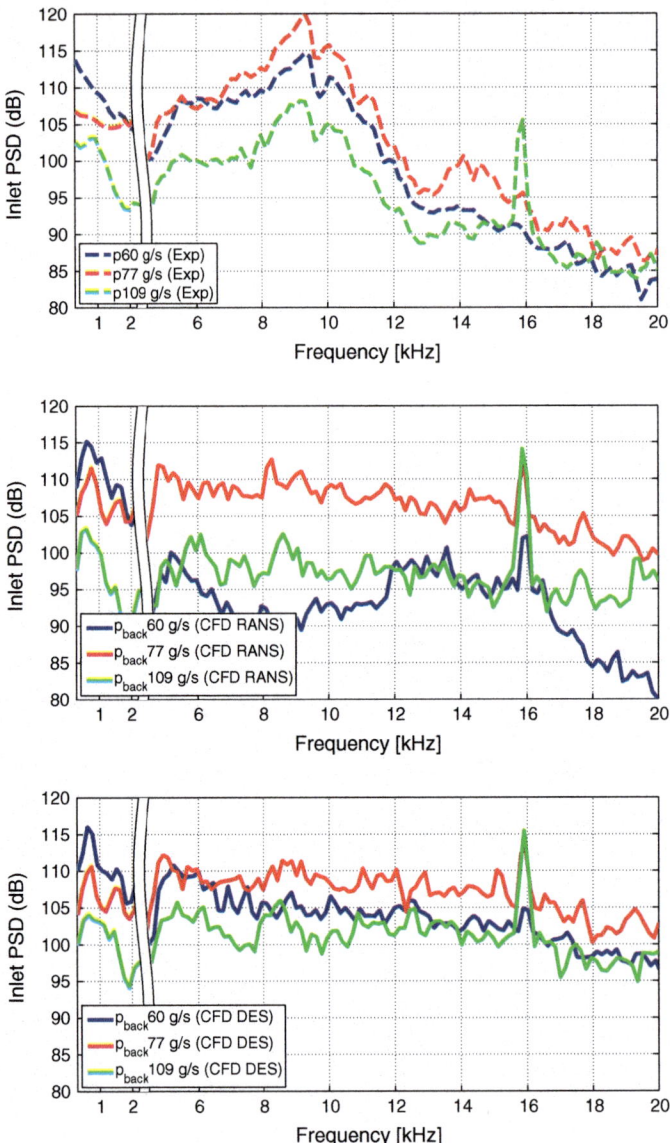

Fig. 6.3 High frequency PSD of experimental pressure (top) and numerical URANS (middle) and DES (bottom) pressure components at inlet duct, for the different operating conditions (reprinted from [6], with permission from Elsevier)

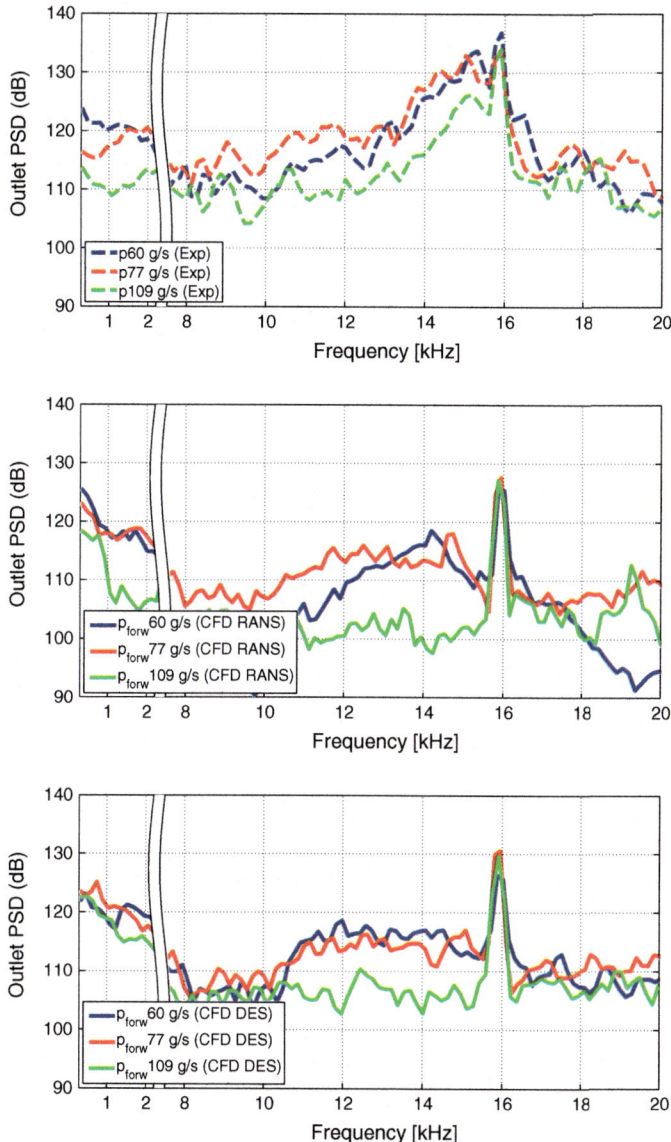

Fig. 6.4 High frequency PSD of experimental pressure (top) and numerical URANS (middle) and DES (bottom) pressure components at outlet duct, for the different operating conditions (reprinted from [6], with permission from Elsevier)

found to be an indicator of whoosh noise onset (see Sect. 2.2.6). Gao et al. [10] did find subsinchronous content attributed to stall in their experimental spectrograms, but at frequencies from 10 to 30% of rotation speed, which may be shifted due to the greater diameter (>0.1 m) of their centrifugal compressor. Similarly, Fujisawa and Ohta [11] found narrow band elevations between 25 and 55% rotation orders in their simulations of a centrifugal compressor employed for marine diesel engines.

Experimental inlet spectra at high frequency range (Fig. 6.3) are alike for all operating conditions: they consist of a broadband noise elevation from 4.7 to 12.7 kHz followed by a decreasing PSD. The differences are that the broadband elevation presents lower amplitude at 109 g/s, a small hump appears between 13 and 15 kHz for 77 g/s, and BPF tone is present near BEP (109 g/s). Numerical simulations predict neither these broadband elevations at 4.7 to 12.7 kHz nor a decreasing PSD at frequencies above 13 kHz. For mass flow rates 77 and 109 g/s, numerical spectra are alike regardless of the turbulence model: flat PSD with the presence of some humps, particularly at 5 and 8.5 kHz. At 109 g/s, DES predicted amplitude is about 5 dB higher in average than URANS PSD, whereas 77 g/s spectra are almost identical for both turbulence approaches. Conversely, at 60 g/s URANS provides a PSD that is below 109 g/s amplitude outside the range of 12 to 17 kHz. It should be noted that, as in the experimental spectra, BPF tone stands out in the inlet numerical PSD near BEP.

Outlet duct spectra at high frequency range (Fig. 6.4) presents a similar description than inlet duct PSD Fig. 6.3. In this case, BPF tones are evident for all operating conditions and experimental broadband elevation ranges from 13 to 16 kHz. Agreement between numerical simulations and measurements is much closer than with inlet spectra, even though the model is unable to provide the broadband elevation from 13 to 16 kHz. Instead, DES does predict a broadband elevation, but from 10 kHz onwards. As for the inlet duct, PSD does not depend on turbulence model for operating 77 and 109 g/s. Close to surge, URANS amplitude is again lower than expected below 12 kHz and above 17 kHz.

6.4 Flow Field Investigation

The flow field provided by the simulations is analyzed in order to find aeroacoustic phenomena that could be the source of the features observed in the PSD in Sect. 6.3. To do so, blade-to-blade surfaces at 50% span will be used for each operating condition. Section A.2.2 describes the obtaining of this postprocessing surface, that starts at the impeller's eye plane and finishes at the volute (see Fig. A.5). In the following analysis, meridional velocity contours are considered instead of pressure ones, since low frequency pressure waves are hidden under BPF tone, whose amplitude and frequency are greater.

Figure 6.5 shows 4 snapshots of meridional velocity at 50% span surface for 109 g/s. These surfaces of revolution are unwrapped by projecting them on a normalized-meridional versus circumferential plane [12]. The sequence of pictures,

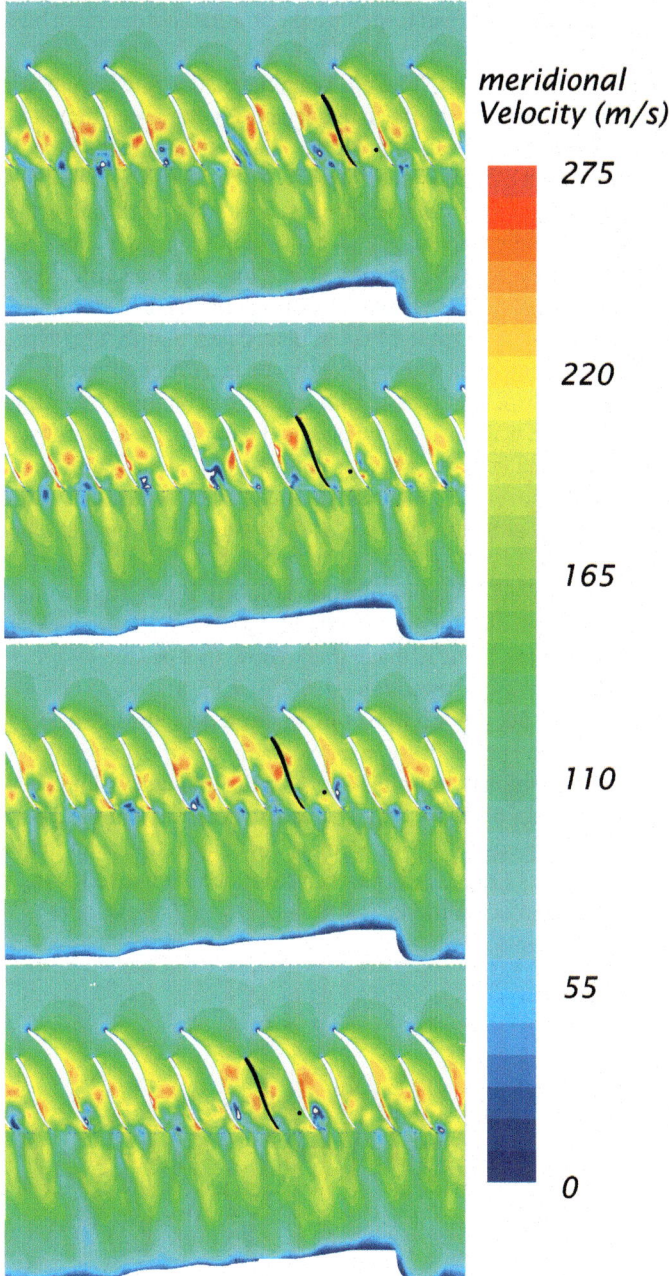

Fig. 6.5 Snapshots of meridional velocity contours at 50% span blade-to-blade surface, for 109 g/s operating point (reprinted from [6], with permission from Elsevier)

representing a time progress if read from top to bottom, covers one main blade passing period.

At these operating conditions, flow upstream the blades is axisymmetric, becoming circumferentially periodic in the inducer and losing any symmetry as stalled regions appear at the exducer of some blades. Earliest flow dettachment occurs at splitter midchord distance. Stall is more intense at the SS of the main blades, which indicates that splitter blades should be closer to main blade PS rather than being a midpitch distance, i.e., upper passages should present a greater cross section than lower passages. Després et al. [9] also noted that lower passages deliver more flow than upper channels for a midpitch location of the splitter blades. Moreover, Jeon [13] noted that, in a centrifugal fan, splitter blades located near main blade PS produce less overall SPL than same impeller with splitter placed at a different pitchwise position.

The stalled regions, which present in some cases recirculating flow (white contours in Fig. 6.5), result in the wake part of the classical jet-wake structure observed at the trailing edge. Meridional velocity displays almost a periodic circumferential profile again at the diffuser. Choi et al. [14] already noticed that a centrifugal pump presents negative radial velocity at blade SS even for operating conditions with higher mass flow than BEP.

Pulsating jets, identified as islands of warm colors in Fig. 6.5 appear at blades SS. Onset of first jets is located upstream stalled regions. These jets are directed towards blades PS as they progress in the exducer due to flow slip, producing pressure waves which travel in the streamwise direction.

The initial (top) and final (bottom) picture of those depicted in Fig. 6.5 show similarities. For instance, the jets at passage corresponding to the black splitter PS are in the same chordwise position at both snapshots, so the frequency of this phenomenon may be close to the main blade passing frequency. Pressure is registered by a point probe located at trailing edge PS, which rotates in conjunction with the impeller as depicted by a black dot in Fig. 6.5. Figure 6.6 shows PSD obtained using this probe for the different operating conditions.

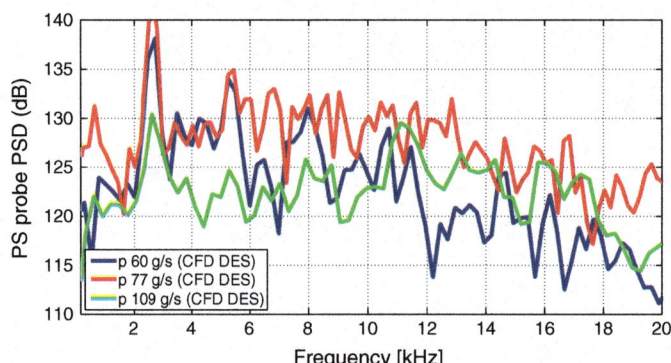

Fig. 6.6 PSD of pressure trace calculated with DES at a probe located at the PS of the traling edge, for different operating conditions (reprinted from [6], with permission from Elsevier)

As will be seen later, the pressure field at the diffuser is strongly unaxisymmetric at 60 and 77 g/s, and thus a tone appears at the rotating order and harmonics. For 109 g/s, this feature also appears, but a narrow band at 11 kHz stands out, which seems to be related with the phenomenon reflected in Fig. 6.5. In a similar way, Bousquet et al. [1] found that 6 vortices are formed and shed per revolution in their centrifugal compressor, thus being responsible of a observed tone at six times compressor rotational frequency. Besides, Pavesi et al. [15] attributed a tone corresponding to 66% of impeller rotation to a sequence of vortices in the wake region originated at blade midchord in a centrifugal pump with vaned diffuser.

Figure 6.7 corresponds to 77 g/s operating conditions. Meridional velocity upstream main blades is almost periodic, except for one stall cell that rotates at about 30% of compressor speed. Bad incidence angle results in flow detachment at main blades leading edge, but passages do not stall until chordwise location corresponding to splitter blades leading edge, further upstream than for 109 g/s mass flow rate. The amount of pitchwise section stalled is increased regarding 109 g/s, and the asymmetry of flow going through main and secondary passages is not as remarkable as in 109 g/s. Flow at the diffuser is no longer axisymmetric: there is a zone of great radial velocity after the tongue followed by a region of stalled flow close to the volute.

Sequence of 4 snapshots included in Fig. 6.7 covers a timespan corresponding to one fourth of a full rotation. Pulsating jets at BPF are not found as in 109 g/s, which is supported by PSD depicted in Fig. 6.6. At 77 g/s, two low frequency phenomena appear. Upstream the impeller, a rotating stall cell has already been identified. At the diffuser and volute, stall cells also rotate at a subsynchronous speed. Regions of flow detached at the blades SS leave the trailing edge with the circumferential velocity imparted by the impeller. Conservation of angular velocity dictates that stalled cells reduce their tangential velocity as they progress in the radial direction through the diffuser. If these stalled cells are tracked, they move one twelfth of a full rotation in the timespan of Fig. 6.7, thus traveling at one third of the rotational speed. Since two stall regions can be identified, this phenomenon would result in an aeroacoustic source of about 1.8 kHz. If a greater timespan is considered, low velocity structures are convected through the outlet duct covering a range of frequencies slightly lower than compressor rotating speed. It is thus proposed that rotating stall is the source of subsinchronous noise known as whoosh noise.

According to Choi et al. [14], jet-wake flow pattern of one passage disturbs adjacent passages, creating a rotating pattern than precesses around the impeller with a speed lower than shaft speed. In this way, even operating conditions not close to surge could present diffuser rotating stall and subsequent whoosh noise, as in the experimental spectrum for 109 g/s (see top of Fig. 6.2).

Fig. 6.8 is the counterpart of Fig. 6.7 but for 60 g/s, i.e., the same timespan is covered by the sequence of snapshots. Flow field at the passages, diffuser and volute looks alike 77 g/s. It agrees with the fact that outlet spectra depicted at Figs. 6.2 and 6.4 for 77 and 60 g/s are very similar but for whoosh noise, which is more remarkable at 60 g/s for DES. The main difference between these two operating points is found upstream the impeller. At 60 g/s, the reversed flow reaches several diameters upstream the impeller whereas for 77 g/s the recirculation does not even reach the impeller's

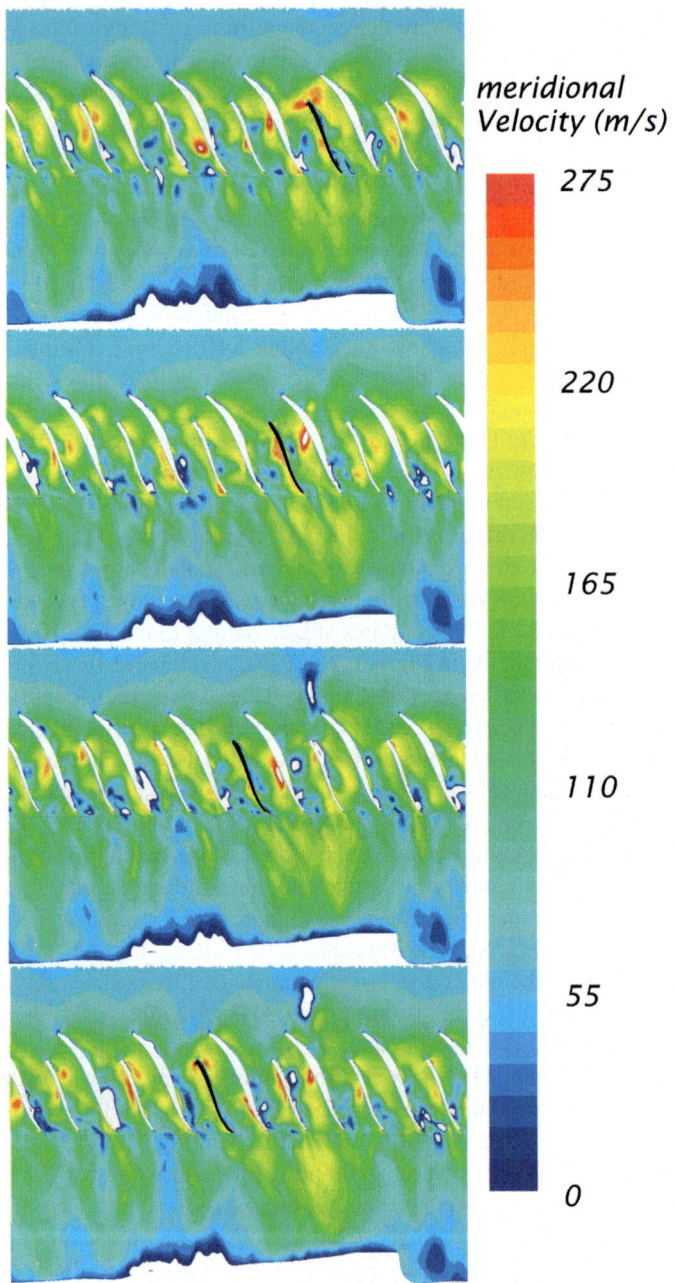

Fig. 6.7 Snapshots of meridional velocity contours at 50% span blade-to-blade surface, for 77 g/s operating point (reprinted from [6], with permission from Elsevier)

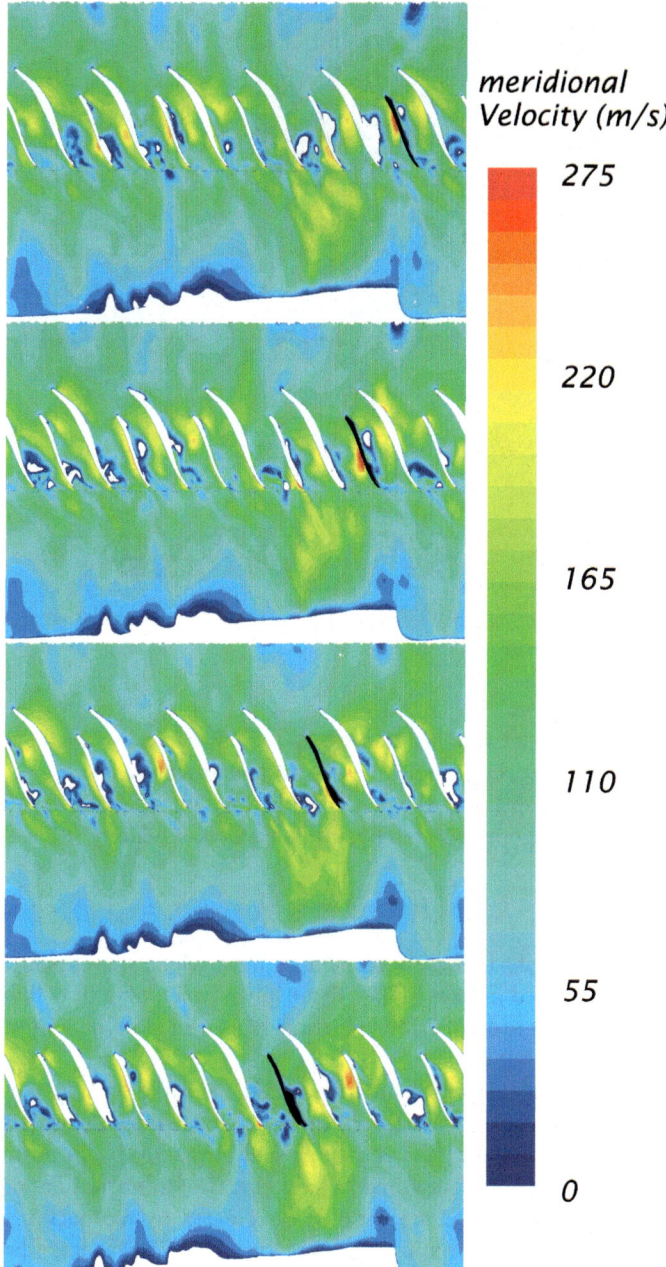

Fig. 6.8 Snapshots of meridional velocity contours at 50% span blade-to-blade surface, for 60 g/s operating point (reprinted from [6], with permission from Elsevier)

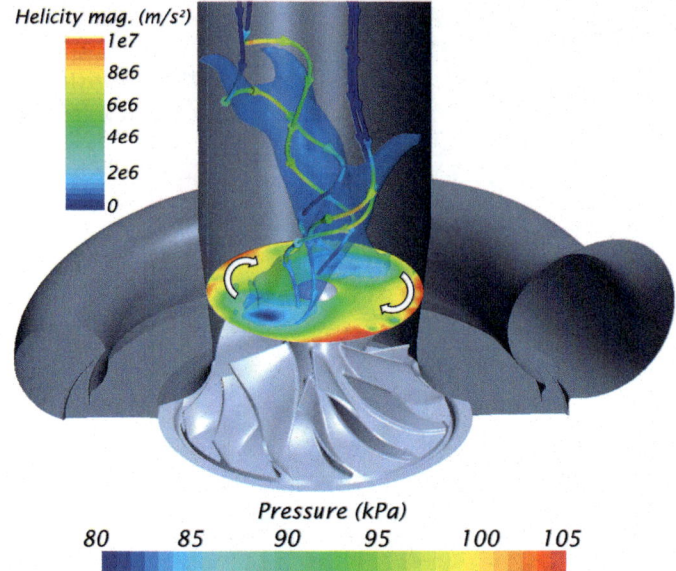

Fig. 6.9 Snapshot of a pair of rotating tornado-type vortices at inlet duct for 60 g/s mass flow rate (reprinted from [6], with permission from Elsevier)

eye. Even though inducer rotating stall exists at both operating conditions, in the case closer to surge the rotating cells present a singular feature.

Fig. 6.9 depicts a snapshot of the flow upstream the impeller at 60 g/s. Incoming streamlines, such as the ones colored by absolute value of helicity,

$$H = \vec{v} \cdot (\nabla \wedge \vec{v}), \tag{6.1}$$

in Fig. 6.9, may suddenly find the recirculating flow, which presents high angular momentum (as discussed in Sect. 3.6). If that happens, incoming flow rolls up, thus increasing vorticity aligned with flow direction. This tornado-type vortex creates a region of low pressure.

For instance, a threshold of pressure below 85 kPa is depicted in transparent blue in Fig. 6.9, showing two *low-pressure bubbles* associated to the vortices. These structures rotate coherently with the inducer rotating stall, as marked by the arrows located over the cross-section which is colored by pressure. The tornadoes are not aligned with the impeller axis because the shear stress created by the backflows induce positive pre-rotation to the incoming flow. The case with 77 g/s cannot present this phenomenon because recirculating flow does not go as far upstream.

Mendonça et al. [2] considered that inducer rotating stall was responsible for the numerical narrow band noise detected at a frequency of 70% of rotational speed. For the compressor and the rotational speed studied in this work, maximum compression ratio point sets the onset of both inducer and diffuser rotating stall, along with whoosh noise. When mass flow is further reduced, inducer rotating stall is accompanied by a rotating low-pressure region created by tornado-type vortices. This phenomenon may increase the PSD below 1.5 kHz seen in Figs. 6.1 and 6.2.

6.5 Conclusions

In this chapter, URANS and detached eddy simulations of a turbocharger compressor at different operating conditions have been performed. Three point have been studied at the same isospeed: one close to the best efficiency point (109 g/s), other at maximum pressure ratio (77 g/s) and one close to surge (60 g/s).

Pressure spectra have been investigated. Experimental PSD show that quietest point is the one with highest mass flow. The other two operating conditions present similar spectra, although point closest to surge is noisier below 1500 Hz. Spectra predicted by simulations display the same tendencies, save for URANS-calculated point at mass flow rate of 60 g/s. *Whoosh noise* is found as a broadband elevation in the range of 1–2.5 kHz. The only experimental PSD feature that simulations cannot reproduce is a broadband noise that appears at the inlet spectra between 4.7 and 12.7 kHz.

Numerical flow field at 109 g/s depicts flow detachment at blades SS. The jet-wake pattern is found to be pulsating at a frequency slightly lower than main BPF. In any case, the flow is quite axisymmetric at these conditions. The other two points present a pattern similar between them and quite different from the one seen at 109 g/s. At the inducer, stall cells are rotating below compressor speed. Moreover, a tornado-type vortex causing a low-pressure bubble is also rotating at the inducer at 60 g/s, thus increasing low frequency noise content. Flow detachment at blades SS is more pronounced than at 109 g/s. Circumferential profile of radial velocity at the diffuser is no longer axisymmetric, being dominated by a region with high speed near the volute tongue. Rotating instabilities are convected to the outlet duct at a frequency in the range of woosh noise. Therefore, it is the proposed fluid mechanism for this acoustic phenomenon.

References

1. Y. Bousquet, X. Carbonneau, G. Dufour, N. Binder, I. Trebinjac, Analysis of the unsteady flow field in a centrifugal compressor from peak efficiency to near stall with full-annulus simulations. Int. J. Rotating Mach. **2014**, 11 (2014). https://doi.org/10.1155/2014/729629
2. F. Mendonça, O. Baris, G. Capon, Simulation of radial compressor aeroacoustics using CFD, in *Proceedings of ASME Turbo Expo2012. GT2012-70028. ASME. 2012*, pp. 1823–1832. https://doi.org/10.1115/GT2012-70028
3. A. Karim, K. Miazgowicz, B. Lizotte, A. Zouani, Computational aero-acoustics simulation of compressor whoosh noise in automotive turbochargers. SAE Technical Paper 2013-01-1880 (2013). https://doi.org/10.4271/2013-01-1880

4. Y. Lee, D. Lee, Y. So, D. Chung, Control of airflow noise from diesel engine turbocharger. SAE Technical Paper 2011-01-0933 (2011). https://doi.org/10.4271/2011-01-0933

5. S. Fontanesi, S. Paltrinieri, G. Cantore, CFD analysis of the acoustic behavior of a centrifugal compressor for high performance engine application. Energy Procedia 45(0) (2014). ATI 2013—68th Conference of the Italian Thermal Machines Engineering Association, pp. 759–768. https://doi.org/10.1016/j.egypro.2014.01.081

6. A. Broatch, J. Galindo, R. Navarro, J. García-Tíscar, Numerical and experimental analysis of automotive turbocharger compressor aeroacoustics at different operating conditions. Int. J. Heat Fluid Flow (2016). https://doi.org/10.1016/j.ijheatfluidflow.2016.04.003

7. L. Mongeau, D. Thompson, D. McLaughlin, Sound generation by rotating stall in centrifugal turbomachines. J. Sound Vib. 163(1), 1–30 (1993). https://doi.org/10.1006/jsvi.1993.1145

8. D. Evans, A. Ward, Minimizing turbocharger whoosh noise for diesel powertrains. SAE Technical Paper 2005-01-2485 (2005). https://doi.org/10.4271/2005-01-2485

9. G. Després, G.N. Boum, F. Leboeuf, D. Chalet, P. Chesse, A. Lefebvre, Simulation of near surge instabilities onset in a turbocharger compressor. Proc. Inst. Mech. Eng., Part A: J. Power Energy 227(6), 665–673 (2013). https://doi.org/10.1177/0957650913495537

10. C. Gao, W. Huang, T. Zheng, K. Yang, H. Yang, Y. Cao, Experimental and numerical investigations of surge extension on a centrifugal compressor with vaned diffuser using steam injection. Int. J. Rotating Mach. 2017, 16 (2017). https://doi.org/10.1155/2017/9159516

11. N. Fujisawa, Y. Ohta, Transition process from diffuser stall to stage stall in a centrifugal compressor with a vaned diffuser. Int. J. Rotating Mach. 2017, 13 (2017). https://doi.org/10.1155/2017/2861257

12. M. Drela, H. Youngren, A user's guide to MISES 2.63. MIT Aerospace Computational Design Laboratory, 2008

13. W.-H. Jeon, A numerical study on the acoustic characteristics of a centrifugal impeller with a splitter. GESTS Int. Trans. Comput. Sci. Eng. 20(1), 17–28 (2005)

14. J.-S. Choi, D.K. McLaughlin, D.E. Thompson, Experiments on the unsteady flow field and noise generation in a centrifugal pump impeller. J. Sound Vib. 263(3), 493–514 (2003). https://doi.org/10.1016/S0022-460X(02)01061-1

15. G. Pavesi, G. Cavazzini, G. Ardizzon, Time-frequency characterization of rotating instabilities in a centrifugal pump with a vaned diffuser. Int. J. Rotating Mach. 2008, 10 (2008). https://doi.org/10.1155/2008/202179

Chapter 7
Concluding Remarks

7.1 Introduction

In this chapter, a critical review of the book is performed. Main findings and contributions of this work are included in Sect. 7.2. Section 7.3 is devoted to the enumeration of limitations of the model and the approach followed in this book, which could affect the validity of the aforementioned findings. Finally, Sect. 7.4 covers potential research topics that may be explored to improve the knowledge of centrifugal compressor aeroacoustics.

7.2 Summary of Findings and Contributions

7.2.1 Main Original Contributions

One of the requirements of any doctoral thesis is that it should include an original contribution that expands the frontiers of knowledge [1]. In this section, the main original contributions of the book are outlined.

- Rotating stall is found to be the underlying mechanism of whoosh noise, detected as a broadband elevation in the range of 1–2.5 Hz, dismissing the role of turbulence that has been claimed by several researchers [2–5]. It should be noted that Mendonça et al. [6] found a subsynchronous narrow band noise which was attributed to rotating stall. However, Mendonça et al. only analyzed inducer rotating stall, and no experimental results were used to confirm that the narrow band noise detected in the simulation corresponded to the phenomenon known as whoosh noise.
- Tip clearance ratio affects compressor global variables (as covered by several researchers [7, 8]), but does not have a significant impact on compressor noise generation, because a coherent noise source cannot be established due to the

© Springer International Publishing AG 2018
R. Navarro García, *Predicting Flow-Induced Acoustics at Near-Stall Conditions in an Automotive Turbocharger Compressor*, Springer Theses,
https://doi.org/10.1007/978-3-319-72248-1_7

recirculating flow. Therefore, CAD tip clearance profile can be safely assumed when simulating centrifugal compressor aeroacoustics.

• Numerical spectra obtained with pressure decomposition based on the Method of Characteristics is in agreement with experimental measurements. For low frequency range (before onset of first asymmetric mode), cross-section monitors provide most consistent PSD. At high frequency, wall point monitors are required to include the effect of higher-order modes.

• The configuration used by Mendonça et al. [6] in terms of time-step size and grid density is found to be appropriate for centrifugal compressor aeroacustic predictions, but now the sensitivity of compressor performance parameters and noise production to these options is known.

  – Using a mesh with 4 million cells deteriorates the prediction of compressor global variables and, over all, noise generation. Conversely, a 19 million elements grid is not found to significantly improve these predictions when compared with a mesh with half the cells. Therefore, a 9 million grid is found to be a good trade-off.

  – At near-surge conditions, segregated and coupled solver provide similar compressor global variables. PSD predictions are in agreement save for inlet duct spectra at high frequency. The segregated solver is selected due to its lower RAM demand.

  – Simulations with $\Delta t = 1°$ are required to maintain accuracy in PSD predictions along all human hearing range. However, the use of $\Delta t = 4°$ still provides adequate spectra in the low frequency range. Compressor performance variables can be predicted with relative difference less than 1% with any of the time-step increments studied, i.e., $\Delta t = 0.5°–4°$.

7.2.2 Other Findings

Other findings that are either a confirmation of past research or are have less impact than the contributions described in Sect. 7.2.1 are the following:

• The quietest point is the one with highest mass flow rate (109 g/s). The other two operating conditions (77 and 60 g/s) present similar spectra, although point closest to surge is noisier below 1500 Hz.

  – Particularly, whoosh noise is increased as mass flow rate is reduced and compressor speed is raised. In a isospeed line, whoosh noise growth is much steeper where pressure ratio decreases with mass flow. In any case, whoosh noise is more intense at the outlet duct.

  – At 60 g/s, recirculating flow creates tornado-type vortices that cause low-pressure bubbles which rotates at the inducer. Low frequency noise content increase may be related to this phenomenon.

- The simulations are able to predict compressor performance variables (pressure ratio, specific work and isentropic efficiency) with a relative difference below 3% for the studied operating conditions.
- DES predicts spectra with slightly better agreement than URANS, particularly at high frequency range (above onset of higher order acoustic modes). Moreover, DES has shown to be less sensitive to mesh density.
- Near surge, the flow in the vicinity of the shroud is reversed. Thus, leakage flow is not only driven by pressure gradient and even there exists zones in which flow goes from blade suction side to the pressure side, due to the angular momentum of the recirculating flow.
- Thermal and rotational effects combined with shaft motion are not expected to reduce CAD tip clearance below 50%.
- At 109 g/s, flow is more *regular* than at the other two working points, i.e., secondary flows are less intense, the flow is more axisymmetric at the diffuser, etc. This case presents recirculating flow, although it is confined to the impeller main passage.

  – Numerical flow field at 109 g/s depicts flow detachment at blades SS. The jet-wake pattern is found to be pulsating at a frequency about 11 kHz.

- Flow at the other two working points (77 and 60 g/s) is alike. This is because case with less mass flow presents thicker backflow zone near the walls, thus presenting a flow similar to the one delivered by 77 g/s case at compressor core. Particularly, inlet recirculation reaches the impeller's eye in 77 g/s, but extends 1 diameter upstream the impeller's eye for 60 g/s.

  – The flow at the diffuser for the two working points with less mass flow rate is clearly not axisymmetric, with a high speed region near the volute tongue.

- Volute is dominated by swirling flow at 109 g/s, which degrades at 77 g/s and is about to break down for 60 g/s.
- Main and secondary passages are not evenly loaded, as found by Després et al. [9]. Therefore, midpitch location of splitter blade seems not to be the best choice for operating conditions in the surge side of compressor map.

7.3 Limitations

As far as the author is concerned, all the objectives stated in Chap. 1 have been accomplished to a great degree, as summarized in Sect. 7.2. However, there should be noted that most of the conclusions obtained in this work are case-dependent. Moreover, the limitations of the numerical model should be taken into account when interpreting the results obtained. In any case, the general agreement between numerical predictions and experimental measurements in terms of global variables and spectra support the validity of the results.

The main limitation of this book is that only one centrifugal compressor has been studied. Impeller geometry, trim, diffuser radial length and volute design are, among others, several factors that shall affect compressor mean flow and aeroacoustics. The conclusions included in Sect. 7.2 are strictly limited to the studied centrifugal compressor.

Besides, only three operating conditions at same compressor speed (160 krpm) have been investigated.

The setup analysis is performed at a unique working point (60 g/s at 160 krpm). The conclusions may not hold for different operating conditions. Particularly, mass flow rate (or, equivalently, Mach number) is expected to have an impact on the differences between performance of segregated and coupled solvers. Compressor speed could affect the decision on the amount of degrees rotated per time step. For lower compressor speeds, physical time-step size (in seconds) could be more restrictive. Regarding mesh independence analysis, only one type of elements (polyhedra) has been employed, although several works [10, 11] have claimed their advantage over other cell geometries.

Tip leakage flow was found to be strongly influenced by recirculating flow. For working points with higher mass flow, noise production sensitivity to tip clearance could differ. For instance, at 109 g/s the secondary passage only presents backflows in the exducer. It should be noted though that compressor noise generation is reduced as mass flow rate is increased, so this potential role of tip clearance in noise production is not capital. Besides, sensitivity of noise generation to tip clearance was investigated, but using a constant (steady) reduced clearance. Actually, shaft motion defines a transient modification in tip clearance, which may affect compressor aeroacoustics.

Existence of standing waves in the simulation implies that spectra depend on monitor axial position. Standing waves appear because the modeled domain only includes a fraction of the turbocharger test rig ducts and NRBC are not used. The effect of standing waves has been mitigated with pressure decomposition, but they are still present in the domain.

Moreover, there are some experimental PSD features that the model is not able to reproduce. Numerical spectra do not include the experimental broadband noises detected between 4.7 and 12.7 kHz for the inlet duct and from 13 to 16 kHz at the outlet. The model also fails to predict a decay in amplitude for frequencies above 13 kHz at the inlet duct. Unfortunately, there are no similar studies that can be used to decide whether the model is not accurate modeling the high frequency noise or these broadband elevation appear due to particular features of the experimental apparatus.

Finally, the determination of rotating stall as the source of whoosh noise is based on observation of velocity contours and the absence of any other phenomena in the same frequency range as whoosh noise. However, pressure fluctuations related to rotating stall are difficult to be appreciated due to two factors. First, blade passing tone presents more amplitude and higher frequency than whoosh noise, thus masking it. Second, pressure rise in the diffuser is more intense than transient pressure fluctuations. The latter effect could have been erased by defining an acoustic pressure in which time-averaged pressure is subtracted from instantaneous one. Nevertheless, this is not suitable because, in this work, the diffuser belongs to the rotating region.

7.4 Suggestions for Future Studies

In this section, suggestion for future studies are proposed. First, natural continuation of research are covered in Sect. 7.4.1, including those topics that could have been identified as interesting beforehand, but were not tackled because the book cannot deal with all the potential research issues due to temporal limitations. Finally, Sect. 7.4.2 includes suggestions motivated by new questions arisen during this work or approaches to overcome the limitations described in Sect. 7.3.

7.4.1 Continuation of Research

Different compressors could be studied to analyze the influence of the geometry in the aeroacoustic phenomena. The approach of Tomita et al. [12] could be followed, selecting compressors with marked differences in acoustic signature in order to simulate the unsteady fluid flow and thus understand the correlation between compressor geometry, transient fluid phenomena and noise generation.

Simulations at other operating conditions should be performed in order to complete the portrait of compressor acoustic signature. At least, points with same relative surge margin but different compressor speeds should be calculated to investigate whether the aeroacoustic phenomena observed in this paper appears at different speeds. Isospeed points in the choke side could also be studied to shed some light on compressor flow-induced acoustics at high mass flow conditions.

Previous experience [13–16] shows that compressor inlet conditions have a great influence on surge onset. It seems interesting then to analyze the impact of inlet geometries on compressor noise generation.

The book has covered only internal flow acoustics, but the noise is mainly appreciated through vibration and subsequent radiation of inlet and outlet hoses. Compressor piping aerovibroacoustics should also be investigated to complete the acoustic transmission path.

7.4.2 Overcoming Thesis Limitations and New Questions Arisen During the Thesis

Inlet NRBC could have been used with a not so great increase of computational effort. In any case, this change of boundary condition does not affect outlet spectra, which are more important due to their higher amplitude. At the outlet duct, the use of NRBC lead the compressor into surge. It should be studied how to improve the performance of such a boundary condition in the outlet. The implementation of radial equilibrium in the NRBC (see the work of Galindo et al. [17]) or a longer extrusion of the outlet duct could be the first hints.

The experimental validation could be further extended by using flush mounted pressure transducers *inside* the compressor. The challenge is even greater than the one faced by Raitor and Neise [18] due to the reduced size of the compressor, but it should be worth it.

Moreover, the broadband noise elevations detected at inlet and outlet experimental spectra should be further investigated in terms of underlying flow mechanism and dependence on operating conditions.

The potential influence of shaft motion in compressor aeroacoustics should be studied. The problem is that rigid body motion cannot be used because the precession pattern that misaligns the shaft and the axis of revolution of the compressor. A dynamic mesh approach should be used instead, thus increasing the computational effort.

The outlet sliding interface should be set in the impeller outlet so as to allow the obtaining of time-averaged values in the diffuser in an absolute frame.

A local mesh independence as performed by Galindo et al. [19] but focusing on PSD prediction seems quite interesting. In this way, not only a mesh distribution close to the optimal one would be obtained but also the importance of each compressor zone in noise production could be estimated. Moreover, other grid elements could be investigated (such as hexaedra) and a finer mesh should be simulated to confirm that the noise PSD converge to the ones showed in this book.

The sensitivity of the rest of relevant numerical parameters should also be considered, particularly for IDDES simulations because there is no much related know-how yet. For this turbulence model, more sanity checks should be performed to assess the resolved turbulent scales and the possibility of spurious effects such as log-layer mismatch. The discretization scheme or the inlet turbulent boundary conditions could be a first step. Turbulence intensity is not the same when the compressor is mounted on a test rig than when it is working in an ICE. For instance, Zhong [20] investigated the influence of inlet turbulence intensity on axial fan pressure spectra. The overall noise increased with turbulence intensity, being this change more sharp below $I = 15\%$.

The identification of aeroacoustic phenomena in this work has been based on the observation of transient evolution of the flow. However, flow decomposition methods such as Dynamic Mode Decomposition (DMD) or Proper Orthogonal Decomposition (POD) could have been used to identify flow structures attributed to certain frequencies. Researchers such as Alenius [21], Sakowitz et al. [22] and Kalpakli et al. [23] have shown the validity of these techniques to extract coherent flow structures in engine-like flow conditions.

References

1. M.D.E.Y. CIENCIA, REAL DECRETO 1393/2007, de 29 de octubre, por el que se establece la ordenación de las enseñanzas universitarias oficiales. *Boletín Oficial del Estado* 260 (2007), pp. 44037–44048
2. D. Evans, A. Ward, Minimizing turbocharger whoosh noise for diesel powertrains. SAE Technical Paper. 2005-01-2485 (2005). https://doi.org/10.4271/2005-01-2485
3. C. Teng, S. Homco, Investigation of compressor whoosh noise in automotive turbochargers. SAE Int. J. Passeng. Cars-Mech. Syst. **2**(1) (2009), pp. 1345–1351. https://doi.org/10.4271/2009-01-2053
4. G. Gaudé, T. Lefèvre, R. Tanna, K. Jin, T. J. B. McKitterick, S. Armenio, Experimental and computational challenges in the quantification of turbocharger vibro-acoustic sources, in *Proceedings of the 37th International Congress and Exposition on Noise Control Engineering 2008*, vol. 2008. 3. Institute of Noise Control Engineering. 2008, pp. 5598–5611. ISBN: 978-1-60560-989-8
5. C. Sevginer, M. Arslan, N. Sonmez, S. Yilmaz, Investigation of turbocharger related whoosh and air blow noise in a diesel powertrain, in *Proceedings of the 36th International Congress and Exposition on Noise Control Engineering 2007*, 2007, pp. 476–485. ISBN: 978-1-60560-385-8
6. F. Mendonça, O. Baris, G. Capon, Simulation of radial compressor aeroacoustics using CFD, in *Proceedings of ASME Turbo Expo 2012*. GT2012-70028. ASME. 2012, pp. 1823–1832. https://doi.org/10.1115/GT2012-70028
7. S.N. Danish, M. Chaochen, Y. Ce, L. Wei, Comparison of two methods to increase the tip clearance and its effect on performance of a turbocharger centrifugal compressor stage, in *International Conference on Energy and Environment* (2006) p. 7
8. Y. Jung, M. Choi, S. Oh, J. Baek, Effects of a nonuniform tip clearance profile on the performance and flow field in a centrifugal compressor. Int. J. Rotating Mach. **2012** (2012). https://doi.org/10.1155/2012/340439
9. G. Després, G.N. Boum, F. Leboeuf, D. Chalet, P. Chesse, A. Lefebvre, Simulation of near surge instabilities onset in a turbocharger compressor. Proc. Inst. Mech. Eng., Part A: J. Power Energy **227**(6), 665–673 (2013). https://doi.org/10.1177/0957650913495537
10. M. Tritthart, D. Gutknecht, Three-dimensional simulation of free-surface flows using polyhedral finite volumes. Eng. Appl. Comput. Fluid Mech. **1**, 1–14 (2007)
11. O. Baris, F. Mendonça, Automotive turbocharger compressor CFD and extension towards incorporating installation effects, in *Proceedings of ASME Turbo Expo 2011: Power for Land, Sea and Air*. ASME, 2011, pp. 2197–2206. https://doi.org/10.1115/GT2011-46796
12. I. Tomita, S. Ibaraki, M. Furukawa, K. Yamada, The effect of tip leakage vortex for operating range enhancement of centrifugal compressor. J. Turbomach. **135**(5), 8 (2013). https://doi.org/10.1115/1.4007894
13. R. Lang, *Contribución a la Mejora del Margen de Bombeo en Compresores Centrífugos de Sobrealimentación*. Ph.D. Thesis. Universitat Politècnica de València, 2011, http://hdl.handle.net/10251/12331
14. J. Galindo, J.R. Serrano, X. Margot, A. Tiseira, N. Schorn, H. Kindl, Potential of flow pre-whirl at the compressor inlet of automotive engine turbochargers to enlarge surge margin and overcome packaging limitations. Int. J. Heat Fluid Flow **28**(3), 374–387 (2007). https://doi.org/10.1016/j.ijheatfluidflow.2006.06.002
15. J. Galindo, F. Arnau, A. Tiseira, R. Lang, H. Lahjaily, T. Gimenes, Measurement and modeling of compressor surge on engine test bench for different intake line configurations. SAE Technical Paper 2011-01-0370 (2011), https://doi.org/10.4271/2011-01-0370
16. J.R. Serrano, X. Margot, A. Tiseira, L.M. García-Cuevas, Optimization of the inlet air line of an automotive turbocharger. Int. J. Engine Res. **14**(1), 92–104 (2013). https://doi.org/10.1177/1468087412449085
17. J. Galindo, A. Tiseira, P. Fajardo, R. Navarro, Analysis of the influence of different real flow effects on computational fluid dynamics boundary conditions based on the method of characteristics, in *Mathematical and Computer Modelling* **57**(7–8) (2013). Public Key Services and

Infrastructures EUROPKI-2010-Mathematical Modelling in Engineering & Human Behaviour 2011, pp. 1957–1964. ISSN: 0895-7177. https://doi.org/10.1016/j.mcm.2012.01.016

18. T. Raitor, W. Neise, Sound generation in centrifugal compressors. J. Sound Vib. **314**, 738–756 (2008). ISSN: 0022-460X. https://doi.org/10.1016/j.jsv.2008.01.034

19. J. Galindo, S. Hoyas, P. Fajardo, R. Navarro, Set-up analysis and optimization of CFD simulations for radial turbines. Eng. Appl. Comput. Fluid Mech. **7**(4), 441–460 (2013). https://doi.org/10.1080/19942060.2013.11015484

20. F.-Y. Zhong, Studies on the aeroacoustics of turbomachinery. J. Therm. Sci. **8**(1), 9–22 (1999). ISSN: 1003-2169. https://doi.org/10.1007/s11630-999-0019-3

21. E. Alenius, *Flow duct acoustics: an LES approach*. Ph.D. Thesis. KTH, MWL Flow acoustics, 2012, pp. viii, 175

22. A. Sakowitz, M. Mihaescu, L. Fuchs, Flow decomposition methods applied to the flow in an IC engine manifold. Appl. Therm. Eng. **65**(1–2), 57–65 (2014). ISSN: 1359-4311. https://doi.org/10.1016/j.applthermaleng.2013.12.082

23. A. Kalpakli, R. Örlü, P. Alfredsson, Vortical patterns in turbulent flow downstream a 90° curved pipe at high Womersley numbers. Int. J. Heat Fluid Flow **44**, 692–699 (2013). https://doi.org/10.1016/j.ijheatfluidflow.2013.09.008

Appendix A
Radial Turbomachinery Postprocessing Tools

A.1 Introduction

This appendix is devoted to the explanation of those postprocessing techniques used throughout the book that are not obtained in a straightforward procedure, using the most up-to-date version of StarCCM+ at the moment of writing the related doctoral thesis, which was 9.02.005 [1]. Section A.2 deals with the definition of postprocessing surfaces that facilitate the understanding of flow behavior in radial turbomachines, and Sect. A.3 presents some field functions that are interesting for the study of compressor aeroacoustics.

A.2 Postprocessing Surfaces

Several CFD codes allow one to create postprocessing surfaces that are specific for radial turbomachinery. However, the codes usually require the definition of a structured domain in the impeller passages, which should be delimited at the inlet and outlet by grid sections (interfaces), thus conditioning the meshing approach. In this section, these postprocessing surfaces are defined using workarounds so as to avoid the meshing limitations.

In the impeller passages, a point R of a passage can be unequivocally defined as $R = R(m, s, p)$, using three coordinates that are characteristic of the turbo topology:

- Meridional coordinate (m), which increases from the inlet to the outlet of the passage.
- Spanwise coordinate (s), which increases from impeller hub to the shroud.
- Pitchwise (also known as azimuthal or circumferential) coordinate (p), which increases from main blade PS to the SS of the next main blade.

© Springer International Publishing AG 2018
R. Navarro García, *Predicting Flow-Induced Acoustics at Near-Stall Conditions in an Automotive Turbocharger Compressor*, Springer Theses, https://doi.org/10.1007/978-3-319-72248-1

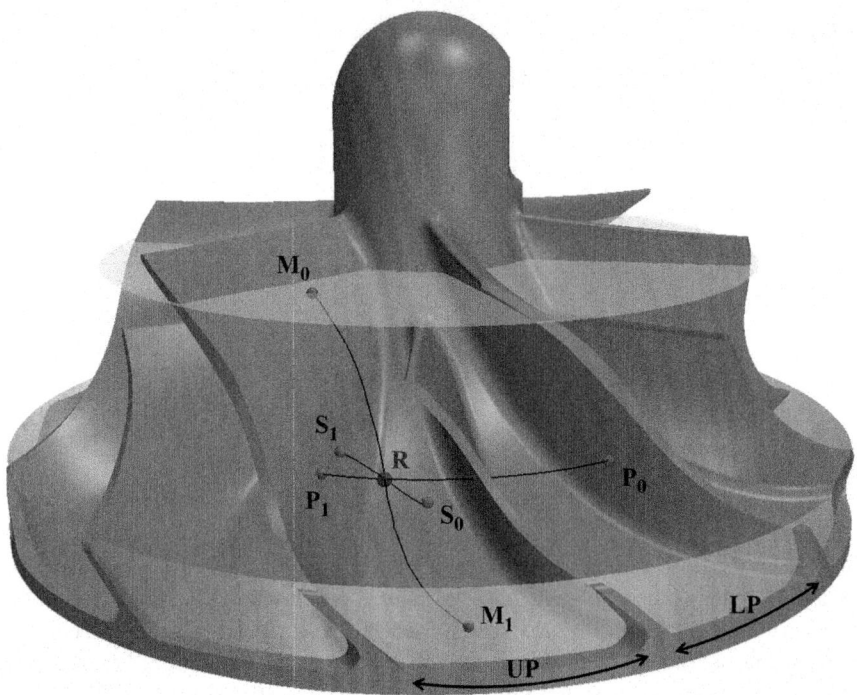

Fig. A.1 Generic point R, located at m = 50%, s = 50% and p = 75%

Following the notation of Fig. A.1, the meridional coordinate is defined as:

$$m(\%) = \frac{\overline{M_0 R}}{\overline{M_0 M_1}} \cdot 100. \qquad (A.1)$$

where M_0 is the point that presents the same spanwise and pitchwise coordinates as point R but is located at the inlet of the passage, i.e., $M_0 = M_0(0, s, p)$, and M_1 is the counterpart of point M_0, being located at the outlet of the passage. Therefore, $M_1 = M_1(1, s, p)$.

Similarly, the spanwise coordinate is defined as:

$$s(\%) = \frac{\overline{S_0 R}}{\overline{S_0 S_1}} \cdot 100. \qquad (A.2)$$

where S_0 is the point that presents the same meridional and pitchwise coordinates as point R but is located at the hub, i.e., $S_0 = S_0(m, 0, p)$, and S_1 is the counterpart of point S_0, being located at the shroud. Therefore, $S_1 = S_1(m, 1, p)$.

Finally, the pitchwise coordinate is defined as:

$$p(\%) = \frac{\overline{P_0 R}}{\overline{P_0 P_1}} \cdot 100. \qquad (A.3)$$

where P_0 is the point that presents the same meridional and spanwise coordinates as point R but is located at the main blade PS, i.e., $P_0 = P_0(m, s, 0)$, and P_1 is the counterpart of point P_0, being located at the SS of the next main blade. Therefore, $P_1 = P_1(m, s, 1)$.

Since two different blades exist (main and splitter blades), two different passages can be distinguished in Fig. A.1:

- Upper passage, delimited by main blade upper side (SS) and splitter blade lower side (PS). Therefore, pitchwise coordinate in the upper passage ranges between $50\% \leq p \leq 100\%$.
- Lower passage, delimited by main blade lower side (PS) and splitter blade upper side (SS). Pitchwise coordinate in the lower passage thus ranges between $0\% \leq p \leq 50\%$.

A.2.1 Isopitch

It is not possible to define a surface with constant pitchwise coordinate analytically, because the shape of the blades is not regular. In this work, only two types of isopitch surfaces are shown:

- Lower passage midpitch surface, which corresponds to p = 25%, thus being located between a main blade PS and a splitter blade SS.
- Upper passage midpitch surface, which corresponds to p = 75%, thus being located between a splitter blade PS and a main blade SS.

In order to obtain these surfaces, a simplified approach is followed by which a main blade PS is rotated 15° and 45° (there are 6 full passages, each of one covering 60°). This method neglects the fact that the blades do not present uniform thickness. However, this error should only be important when defining isopitch surfaces near the blades, which is not the case of this book. Figure A.2 presents the surfaces defined with the aforementioned approach.

Then the surface can be projected onto an axial-radial plane to create the so-called *meridional view*, as can be seen in Fig. A.3. The transformation is provided by StarCCM+ [1].

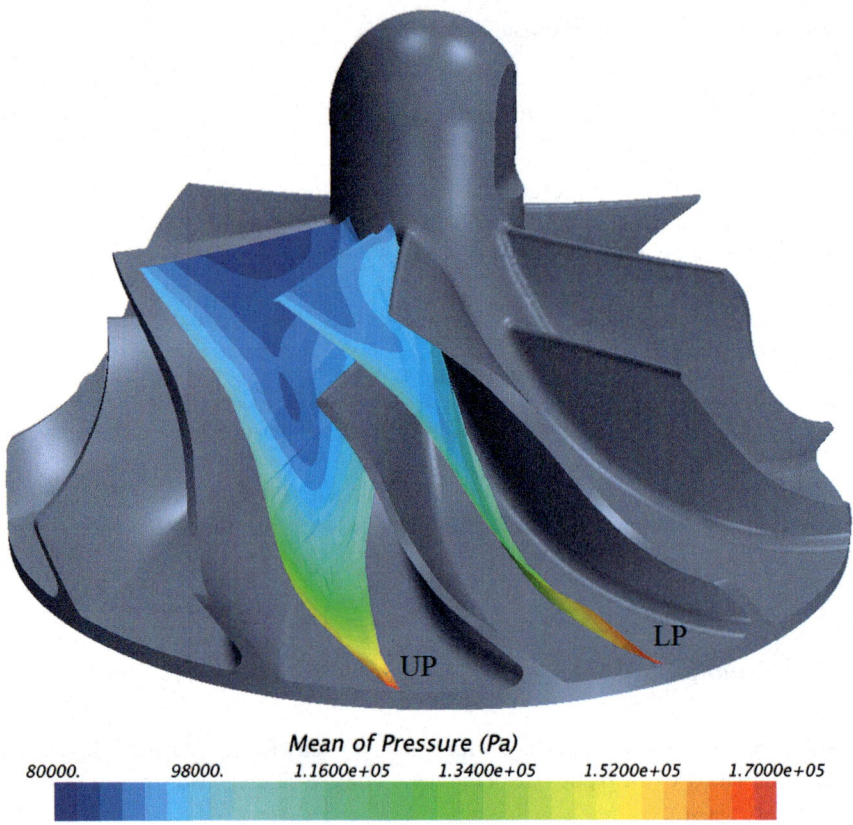

Fig. A.2 3-dimensional view of midpitch surfaces, with contours of mean pressure for 60 g/s mass flow rate

A.2.2 *Isospan*

Figure A.3 suggests that both hub and shroud may be described by an arc of a ellipse. This fact can be used to produce an isospan surface.

Consider Fig. A.4, in which the ellipses for the hub (blue) and the shroud (red) are depicted. Both ellipses are defined using the axial (z) and radial (r) axes, but each of them present different center and semi-axes (Eq. A.4).

$$f(r, z) = \left(\frac{\tilde{r}_i}{a_i}\right)^2 + \left(\frac{\tilde{z}_i}{b_i}\right)^2 = 1, \quad i = \{0(hub), 1(shroud)\} \tag{A.4}$$

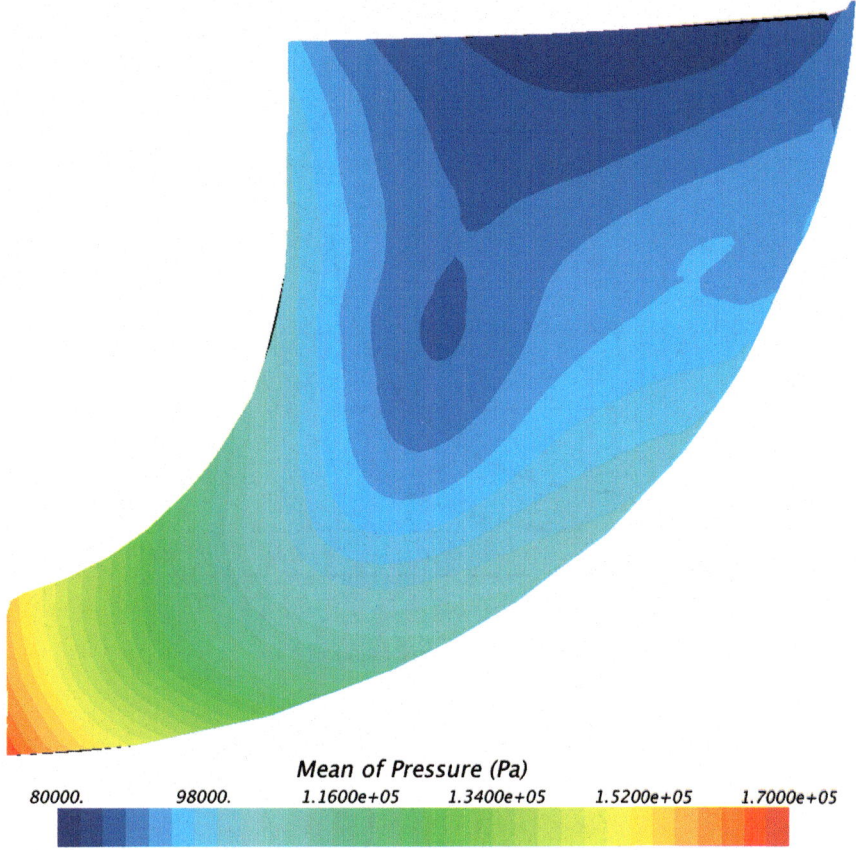

Fig. A.3 Meridional view of main passage midpitch surface, with contours of mean pressure for 60 g/s mass flow rate

where,

$$\widetilde{r}_i = r - r_{c,i}$$
$$a_i = r_{c,i} - r_{a,i}$$
$$\widetilde{z}_i = z - z_{c,i}$$
$$b_i = z_{c,i} - z_{b,i}, \tag{A.5}$$

would describe the ellipse for the hub and the shroud. Then, an isospan is defined as an ellipse whose center and semi-axes change linearly with s(%) from the ones corresponding to the hub (s = 0%) to the ones related to the shroud (s = 100%).

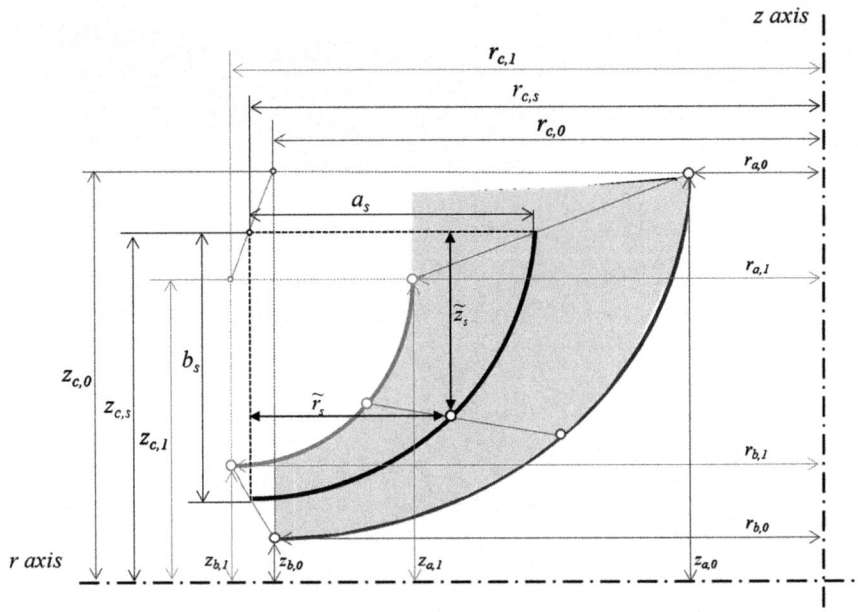

Fig. A.4 Sketch of elliptical representation of isospan surfaces

The approach is shown in Fig. A.4 and reflected in the following equations:

$$f(r, z, s) = \left(\frac{\widetilde{r}_s}{a_s}\right)^2 + \left(\frac{\widetilde{z}_s}{b_s}\right)^2 = 1 \tag{A.6}$$

where,

$$
\begin{aligned}
\widetilde{r}_s &= r - \big[s\, r_{c,1} + (1 - s)r_{c,0}\big] \\
a_s &= \big[s\, r_{c,1} + (1 - s)r_{c,0}\big] - \big[s\, r_{a,1} + (1 - s)r_{a,0}\big] \\
\widetilde{z}_s &= z - \big[s\, z_{c,1} + (1 - s)z_{c,0}\big] \\
b_s &= \big[s\, z_{c,1} + (1 - s)z_{c,0}\big] - \big[s\, z_{b,1} + (1 - s)z_{b,0}\big].
\end{aligned} \tag{A.7}
$$

If one selects a certain span s^*, Eqs. A.6 and A.7 provide an ellipse, generically defined by $f(r, z, s^*) = 1$. Since $f(r, z, s^*) = 1$ holds for any circumferential coordinate θ, the surface of revolution defined by $f(r, z, s^*) = 1$ is an isospan, which is depicted in green in Fig. A.5. In order to extend the isospan surface to the rest of the compressor, Fig. A.5 presents an inlet cylinder (blue) and surfaces of constant height at diffuser (yellow) and volute (red) as well.

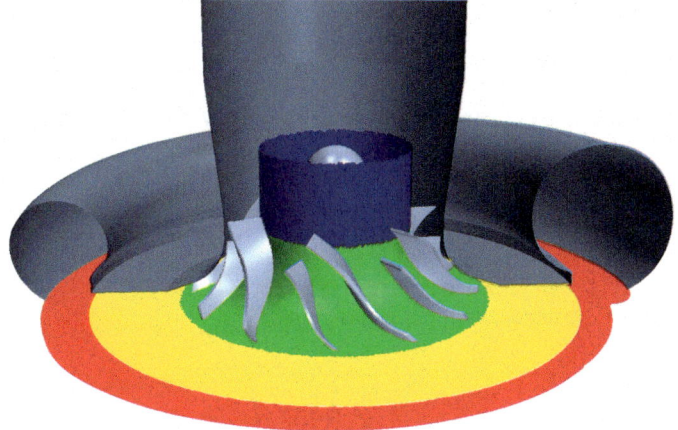

Fig. A.5 3-dimensional view of set of midspan (s = 50%) surfaces

The whole set of surfaces can be unwrapped by projecting them on a normalized-meridional versus circumferential plane [2], which is a feature in StarCCM+. The normalized meridional coordinate is defined by:

$$m' = \int \frac{\sqrt{dr^2 + dz^2}}{r}. \tag{A.8}$$

Figure A.6 shows an example of this flattened view.

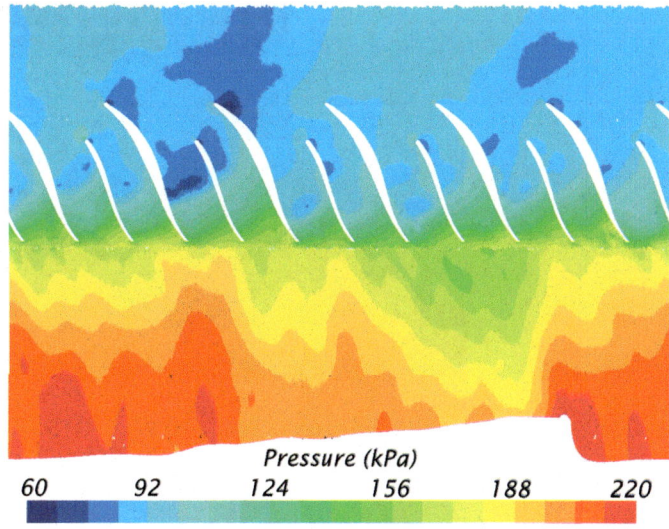

Fig. A.6 Unwrapped view of set of midspan surfaces with contours of pressure

A.2.3   Isomeridional

For a certain span, the ellipse defined by Eq. A.6 can be parametrized using the so-called parametric angle t (also known as the *eccentric anomaly* in celestial mechanics):

$$\tilde{r}_s = a_s \cos t_s$$
$$\tilde{z}_s = b_s \sin t_s \qquad \qquad (A.9)$$

This parametric angle, calculated employing

$$t_s = \arctan\left(\frac{a_s}{b_s}\frac{\tilde{z}_s}{\tilde{r}_s}\right), \qquad \qquad (A.10)$$

would provide a meridional coordinate (see Eq. A.1) for the locus of points at same span s. In order to eliminate the dependence of the span, so that a proper isomeridional surface can be defined, Eq. A.6 should be used to obtain an explicit definition of $s = g(r, z)$. However, this cannot be done analytically, because a complete polynomial of order 4 in s is obtained. A workaround consists in defining the isomeridional surface by setting a certain value of t_s for the most representative value of s, i.e., s = 50%. In this way, Fig. A.7 confirms that this approach is suitable to define proper isomeridional surfaces.

Fig. A.7 Contours of parametric angle $t_{0.5}$ at main passage midpitch surface

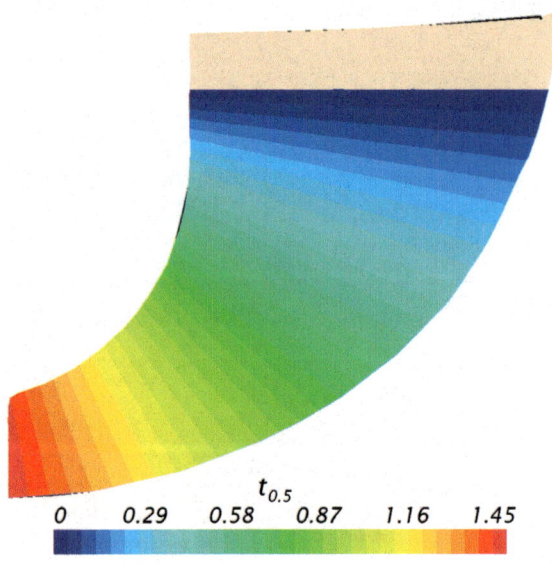

A.3   Field Functions

In addition to the standard field functions offered in StarCCM+ [1], some specific variables are defined in this section that are used throughout the book.

A.3.1   Meridional Velocity

First, it would be interesting to create a scalar field of meridional velocity, so as to identify stalled cells. According to Eq. A.9, a meridional coordinate vector $\vec{m}_s$ could be defined as:

$$\vec{m}_s = a_s \sin t_s \vec{r} - b_s \cos t_s \vec{z}. \tag{A.11}$$

The problem is that StarCCM+ only allows a vector to be defined in the laboratory coordinate system. Hence, the vector should be transformed from the cylindrical coordinate system to a Cartesian one by means of

$$\begin{bmatrix} m_{s,x} \\ m_{s,y} \\ m_{s,z} \end{bmatrix} = \begin{bmatrix} \cos\theta & -\sin\theta & 0 \\ \sin\theta & \cos\theta & 0 \\ 0 & 0 & 1 \end{bmatrix} \begin{bmatrix} m_{s,r} \\ m_{s,\theta} \\ m_{s,z} \end{bmatrix}$$

The corresponding meridional versor $\widehat{m_s}$ is obtained using:

$$\widehat{m_s} = \frac{\vec{m}_s}{|\vec{m}_s|}. \tag{A.12}$$

These versors hold for a certain isospan. Again, 50% span is used to define the versors that are employed in the whole passage. Figure A.8 shows that the approach provides no significant error (the meridional view distorts the actual direction of the versors at the passage outlet).

Finally, the meridional velocity can be derived using

$$v_{m,s} = \vec{v} \cdot \widehat{m_s}. \tag{A.13}$$

Meridional velocity at the rest of the surfaces depicted in Fig. A.5 is obtained as an axial velocity at inlet cylinder, using radial velocity at diffuser and volute surfaces. Figure A.9 presents an unwrapped view of meridional velocity throughout the compressor.

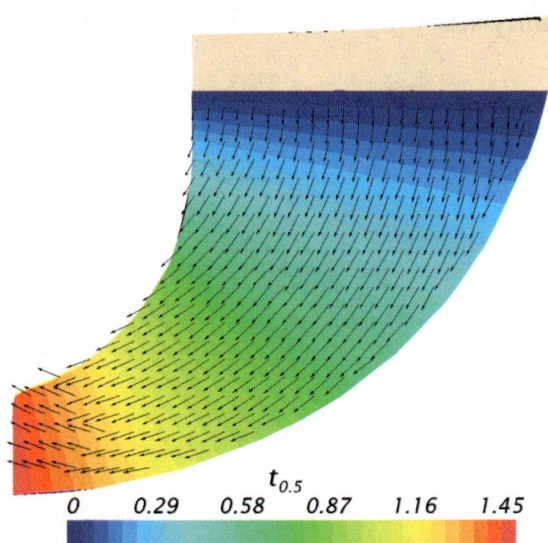

Fig. A.8 Meridional versors $\widehat{m_{0.5}}$ at main passage midpitch surface

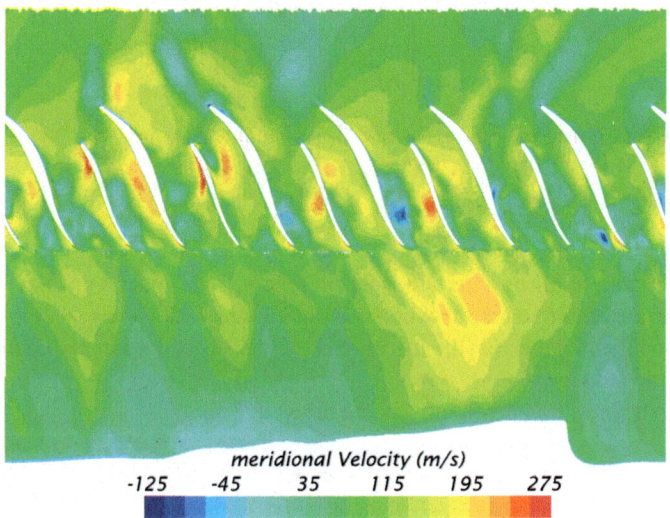

Fig. A.9 Meridional velocity contours at midspan surface

A.3.2 Time-Averaged Field Functions

StarCCM+ offers a tool named as "mean monitor", by which a generic scalar field can be time-averaged. For each cell, an arithmetic mean is performed considering data from the time step n_0 at which the mean monitor is activated until the collection of data is stopped N time steps latter, i.e.,

$$\bar{\phi} = \frac{1}{N} \sum_{n=n_0}^{n_0+N} \phi_n.$$

(A.14)

A time-averaged vector can be obtained by considering the mean monitor of each component. For instance, the time-averaged absolute velocity can be defined as:

$$\bar{\vec{v}} = \overline{v_x}\vec{i} + \overline{v_y}\vec{j} + \overline{v_z}\vec{k}.$$

(A.15)

Mean absolute velocity vector can be used at the stationary regions: inlet duct, volute and outlet duct. However, at rotating regions (rotor and diffuser), a velocity defined using Eq. A.15 would provide spurious "x" and "y" components because each cell is constantly changing these axes according to the rigid body motion. Instead, a cylindrical coordinate system should be used to collect the velocity components $\overline{v_r}$, $\overline{v_\theta}$ and $\overline{v_z}$. Again, the vector should be expressed in the laboratory coordinate system. Therefore,

$$\begin{bmatrix} \overline{v_x} \\ \overline{v_y} \\ \overline{v_z} \end{bmatrix} = \begin{bmatrix} \cos\theta & -\sin\theta & 0 \\ \sin\theta & \cos\theta & 0 \\ 0 & 0 & 1 \end{bmatrix} \begin{bmatrix} \overline{v_r} \\ \overline{v_\theta} \\ \overline{v_z} \end{bmatrix}$$

Once the Cartesian components are known, the absolute velocity in the rotating region can be defined using Eq. A.15. This time-averaged velocity can be straightforward used in the diffuser. In the rotor, it is more appropriate to define a mean relative velocity by subtracting the blade speed:

$$\bar{\vec{v}} = \bar{\vec{v}} - \vec{\omega} \wedge \vec{r}.$$

(A.16)

References

1. STAR-CCM+, Release version 9.02.005. CD-adapco (2014). http://www.cd-adapco.com
2. M. Drela, H. Youngren, *A User's Guide to MISES 2.63* (MIT Aerospace Computational Design Laboratory, Cambridge, 2008)

About the Author

Roberto Navarro was born in Valencia (Spain) in 1986. He obtained his M.Sc. degree in Industrial Engineering in 2009 and his Ph.D. degree in Transport Propulsion Systems in 2014, both at Universitat Politècnica de València, where he is currently employed as a lecturer. Particularly, he is a member of CMT-Motores Térmicos research institute and his area of interest is focused on 3D-CFD simulations of turbochargers. Pulsating flow at turbines, compressor aeroacoustics and surge are the main phenomena that Navarro has analyzed using CFD. Besides, he is involved in many other topics related with engine air management, such as 0D-1D engine modeling or analyzing formation of water droplets and subsequent impact on compressor impeller due to long-route exhaust gas recirculation (LR-EGR). In these fields, Navarro has published 12 scientific papers in JCR journals and 10 contributions to conferences, and has participated in 5 competitive public-funded research programs and 15 private-funded research collaborations. He has received several awards, including an outstanding Ph.D. thesis prize by Universitat Politècnica de València.

© Springer International Publishing AG 2018

R. Navarro García, *Predicting Flow-Induced Acoustics at Near-Stall Conditions in an Automotive Turbocharger Compressor*, Springer Theses, https://doi.org/10.1007/978-3-319-72248-1

Printed by Printforce, the Netherlands